*Sources for the History of
the Science of Steel
1532–1786*

This is Number 4 in a series of monographs in the history of technology and culture published jointly by the Society for the History of Technology and the M.I.T. Press. The members of the editorial board for the Society for the History of Technology Monograph Series are Melvin Kranzberg, Cyril S. Smith, and R. J. Forbes.

Publications in the series include:

History of the Lathe to 1850
Robert S. Woodbury

English Land Measuring to 1800: Instruments and Practices
A. W. Richeson

The Development of Technical Education in France 1500–1850
Frederick B. Artz

Sources for the History of the Science of Steel 1532–1786
Cyril Stanley Smith

Sources for the History of the Science of Steel
1532—1786

Edited by
Cyril Stanley Smith

Published jointly by
The Society for the History of Technology
and
The M.I.T. Press
Massachusetts Institute of Technology
Cambridge, Massachusetts, and London, England

Copyright © 1968 by
The Massachusetts Institute of Technology

Set in Monotype Bembo
Printed by Halliday Lithograph Corporation
Bound in the United States of America by The Colonial Press, Inc.

All rights reserved. No part of this book may be reproduced or utilized in any form or by any means, electronic or mechanical, including photocopying, recording, or by any information storage and retrieval system, without permission in writing from the publisher.

Library of Congress catalog card number: 68-14448

Preface

This collection of papers, the most significant of which appear in English for the first time, is intended to provide the historian, the practical metallurgist, and the student with a glimpse of the rather slow development of an important branch of technology from an art into a science. The discovery of steel occurred over three millenia ago and was not in any way dependent upon theory, but the vast increase in the amount of it and in its reliability, especially in the heat-treated condition, has been in large measure the outcome of the increased understanding that has come about in the last two centuries.[a]

Iron did not take its place among man's materials until somewhat after bronze, and it did not become important until about 1200 B.C. Its processing was unrelated to that of the nonferrous metals, for it was

---

[a] This book is not, of course, meant to stand by itself. For perspective the excellent History of Metals (London, 1960) by Leslie Aitchison should be read. For the early history of steelmaking, Ludwig Beck, Geschichte des Eisens (Braunschweig, 1891–1903), remains the most complete, though Otto Johannsen, Geschichte des Eisens (Düsseldorf, 1953), is more readable and up-to-date. The History of the British Iron and Steel Industry by Hans R. Schubert (London, 1957) is excellent, if geographically limited. For a world view of steel, see The Coming of the Age of Steel by T. A. Wertime, which is erratic in spots but replete with suggestive insights. C. S. Smith's A History of Metallography (Chicago, 1960, reprinted 1965) devotes several chapters to the knowledge of structure before the microscope was used and emphasizes the role that studies of steel played in developing the concepts of the modern science of materials. Finally, for the most recent period, there is the Proceedings of the Sorby Centennial Symposium on the History of Metallurgy, C. S. Smith, editor (New York, 1966). The American Society for Metals has recently announced a plan to publish a series of annotated reprints of "Metallurgical Classics" in the ASM Transactions Quarterly.

forged to shape, not melted, and its superiority over bronze was at first mostly an economic one since its ores were plentiful, cheap, and widely distributed. Steel, which today we know to be essentially an alloy of iron with usually less than one part in a hundred of carbon, was originally thought to be either a different species of material with its own ore, related to iron but different in nature and requiring special metallurgical treatment, or, especially following Aristotle, steel was assumed to be a more pure form of iron[b]*—a reasonable enough assumption since it was produced by prolonged heating in a fire. It was obvious enough that bronze was a compound material, since a particular mixture of ores or metals was necessary to produce it; and even though zinc was not available as a separate metal until the sixteenth century, it was clear that brass resulted from the transfer of some weighable amounts of matter from calamine to copper. Steel, on the other hand, needed no tangible additions—only the carbon that, with the other unsuspected ingredient, oxygen, made the fire. The discovery of the presence of carbon and its chemical role in steel was a triumph of eighteenth-century analytical chemistry, and the principal documents are included here.*

But a full understanding of the almost miraculous changes in the properties of steel as it is heat-treated requires attention to structure as

---

[b] *Aristotle introduced steel in a discussion of softening, liquefaction, and solidification in his* Meteorologica, *Book IV, Chapter 6. The relevant passage is translated by H. D. P. Lee in the Loeb Classical Library edition (London, 1952) as follows:* "Wrought iron indeed will melt and grow soft, and then solidify again. And this is the way in which steel is made. For the dross sinks to the bottom and is removed from below, and by repeated subjection to this treatment the metal is purified and steel produced. They do not repeat the process often, however, because of the great wastage and loss of weight in the iron that is purified. But the better the quality of the iron the smaller the amount of impurity." *This reference to melting has confused generations of scholars, for cast iron was not intentionally produced in Europe much before A.D. 1400 The word* τηκεται *in this instance is usually assumed to have meant softening, not complete liquefaction, but it could have referred to a molten slag, or even to an evanescent formation of molten cast iron, which would have soaked into the sponge iron by capillarity and solidified again as the carbon diffused into the iron to give steel.*

well as to composition. Though appreciated in a qualitative way from the beginning, it was not until well after the discovery of carbon that the microstructural basis of the visual changes in the texture of steel was properly understood. Details of the appearance of fractured surfaces of iron and steel were first used in scientific studies early in the eighteenth century, but it was not until nearly the end of the nineteenth century that the crystalline nature of the artisan's age-old "grain" was widely appreciated. It followed the use of the microscope on polished and etched sections, which revealed hitherto-unsuspected structural details that could be related to the composition, to the treatment, and to the properties of the metal.

The papers reprinted here cover a period of 250 years. Except in the case of the first item, the writers were near the forefront of advancing technology or science, and the ideas expressed do not represent the knowledge of the common artisan. The first four papers are strictly practical, with no shred of theory, yet the increasing precision in the control of practice is obvious. Jousse's studies of fracture in relation to the utility of iron and steel for various purposes (Chapter Four) were followed by those of Réaumur (Chapter Five) who used fracture scientifically as the basis for understanding the differences between steels and to give a structural theory of the mechanism of hardening. This fertile approach was not followed up for over a century, mainly because the Cartesian corpuscular view of the structure of matter that Réaumur expressed was at best qualitative and unverifiable, and it became unpopular among scientists. Physicists concentrated on mechanics and optics, while with chemists composition took the stage as their ancient qualities and principles were vanquished. Metallurgical developments were intimately mixed up with the eighteenth-century attempts to elucidate the role of phlogiston. Chemistry in the eighteenth century was primarily the development of analytical methods and of the realization that there is a relatively small number of elements whose combinations make up all matter. It saw the transition from a concern with qualities believed to be transferable in some degree from substance to substance, to a concern, indeed an overconcern, with composition. The powerful analytical methods that revealed many new elements also disclosed an old one in an unsuspected place—carbon in steel—and if the intellectual impact of this was perhaps less immediate than the discovery of oxygen, it had at least as great economic significance. It is significant that the observation of the carbonaceous residue that was left when steel was dissolved in acids, which nucleated the whole chain of events, arose in the examination of textured metals that were being made commercially to match those long known in the Orient.

An explanation is perhaps needed for the fact that the selections deal mainly with the scientific side of steel. This knowledge had no effect on society until it had been translated into practice, and a cynic could even remark that during the entire period covered by the writings in this volume the practice was essentially independent of science. The eighteenth-century innovations in furnace design and operation that enabled the vast increase in production of iron were the products mainly of the English ironmasters, but it was in Sweden and France that science advanced. The sources for studies of such things are massive furnaces and heavy machinery, and these are more difficult to preserve and to read than are the historians' usual verbal sources. These papers, therefore, represent only a small part of the whole, but they are important because they lie at the somewhat neglected interface between a growing industrial organization on one hand and a burgeoning pure science on the other.

The papers herein contain sections that are difficult to understand and arguments that seem to be extended far beyond need. These, of course, are precisely the parts that represented the advancing edge of knowledge at the time they were written, and it will repay the reader's effort if he places himself in the state of mind where he can share the puzzlement and triumph of the writers.

These translations were prepared mainly because the editor, an indifferent linguist, wanted to know exactly what various writers had said at different times about steel, but he is confident that there are *many others who will enjoy having them in accessible form. Despite the great care that has been taken to temper metallurgical sense with linguistic accuracy and historical caution, it is too much to hope that no errors and misinterpretations have crept into the translations. The serious historian must, as always, make his final interpretations on the basis of the author's own words in the original language, not on a translated version. If, however, the perusal of these research papers of the past helps anyone toward additional insights into the fascinating interrelationship between workman and philosopher, between the applied scientist and the pure, the translators' labor will have been well spent.*

In order to distinguish comments by the editor from those of the original authors, the former are printed throughout in italic type. Footnotes designated by letter are also those of the editor; those of the original authors or previously published translators are numerically designated. The interpolated numbers in square brackets in the text denote the pages in the first edition in the original language.

Cambridge, Massachusetts   Cyril Stanley Smith
June 1967

Acknowledgments

This book would not exist without the help of Anneliese Sisco, J. Patrick Hickey, Pauline Boucher, and Anne S. Denman. Their efforts to produce readings that are both linguistically and scientifically acceptable are deeply appreciated. Pauline Boucher has converted mutilated first drafts into clean typescript and been of help in innumerable other ways.

Thanks are due to the U.S. National Library of Medicine, Bethesda, which provided a photocopy of Bergman's Dissertatio, *and to the Iron and Steel Institute Library (London), the Herzog August Bibliothek in Wolfenbüttel, and the British Museum, which made available microfilms of various editions of the* Stahel und Eysen. *Finally, a grant from the National Science Foundation that has aided both the prosecution of the work and its publication is most gratefully acknowledged.*

C.S.S.

Contents

Preface   v–ix
Cyril Stanley Smith

Plates   xvi

CHAPTER ONE

On Steel and Iron   1–19

The Anonymous booklet, *Von Stahel und Eysen* ... (Nuremberg, 1532). Translated by Anneliese G. Sisco.

Introduction   7
How to Make the Electuaries   8
How Iron Is to Be Hardened   9
Recipes for Hardening Steel   11
How Steel Is to Be Made Soft so that it can be cut   13
Soldering   13
How to Etch into Steel and Iron, or upon armor   16
How to Give Different Metals the Color of Silver or Gold   17

CHAPTER TWO

Vannoccio Biringuccio

Concerning the Method of Making Steel   21–28

Extract from *De la Pirotechnia* (Venice, 1540). Translated ... by C. S. Smith and M. T. Gnudi (New York: © American Institute of Mining, Metallurgical and Petroleum Engineers, 1942), pp. 67–70.

CHAPTER THREE

Giovanni Battista della Porta

Of Tempering Steel 29–39

Extract from *Magiae Naturalis Libri Viginti* (Naples, 1589). Reproduced from the anonymous English translation (London, 1658), pp. 305–12.

The Proeme *32*
That Iron by mixture may be made harder *32*
How Iron will wax soft *33*
The temper of Iron must be used upon soft Irons *33*
How for all mixtures, Iron may be tempered most hard *34*
Liquors that will temper Iron to be exceeding hard *35*
Of the temper of a Tool shall cut a Porphyr Marble Stone *36*
How to grave a Porphyr Marble without an Iron tool *37*
How Iron may be made hot in the fire to be made tractable for works *38*
How Damask Knives may be made *38*
How polished Iron may be preserved from rust *39*

CHAPTER FOUR

Mathurin Jousse

The Selection of Iron and Steel, and the Manner of Hardening It by Quenching 41–62

Extract from *La Fidelle Ouverture de l'Art de Serrurier* (La Flêche, 1627). Reprinted with minor changes in footnotes from the English translation by C. S. Smith and A. G. Sisco, *Technology and Culture*, 1961, 2, pp. 131–45 (with permission of Society for the History of Technology).

How to Recognize Iron That is Soft and Can be Worked Cold *44*
How to Distinguish Good Iron from Bad after It Has Been Broken *47*
How to Recognize Iron That is Brittle When Cold *48*
How to Recognize Hot-Short Iron Or Iron That Is Brittle When Hot *48*
How to Distinguish Good Steel from Bad *49*
Piedmont Steel *50*
Another Kind of Piedmont Steel *51*
Steel from Germany *52*

Carme Steel, or Steel with a Rose   52
Steel from Spain   53
Another Steel from Spain Called "Grain" Steel   53
Dealing with Different Methods of Quenching Steel   55
How to Quench the Little Squares of Limousin, Clamecy, and Artificial Steel   56
How to Quench Piedmont Steel   56
How to Quench Spring Steel from Germany   57
How to Quench Carme Steel, or Steel with a Rose   58
How to Quench Steel from Spain   59
How to Harden Files and Other Tools Made of Either Iron or Steel   60
Another Way of Hardening Small Files, Taps, or Drawplates   62

CHAPTER FIVE

René Antoine Ferchault de Réaumur

On Methods of Recognizing Defects and Good Quality in Steel and on Several Ways of Comparing Different Grades of Steel   63–106

Extract from *L'Art de Convertir le Fer Forgé en Acier* ... (Paris, 1722). Reprinted from the English translation by A. G. Sisco (Chicago, 1956), pp. 176–204.

CHAPTER SIX

Johann Andreas Cramer

The Preparation of Steel out of Iron by Cementation and by Fusion   107–117

Extract from *Elementa Artis Docimasticae* (Leyden, 1739). Reproduced from the English translation by Cromwell Mortimer (London, 1741), pp. 344–50.

Apparatus   111
The Use and Reasons of the Process   113
The Production of Steel out of crude unmalleable Iron, or out of its Ore, by Fusion   115

CONTENTS

CHAPTER SEVEN

Pierre Clément Grignon

On the Metamorphoses of Iron   119–164

Extract from the *Mémoires de Physique sur l'Art de Fabriquer le Fer* . . . (Paris, 1775), pp. 56–90. Translated by P. Boucher and C. S. Smith.

CHAPTER EIGHT

Torbern Bergman

A Chemical Essay on the Analysis of Iron   165–255

Dissertatio Chemica de Analysi Ferri (Uppsala, 1781). Translated by J. P. Hickey and C. S. Smith.

The Varieties of Iron   *172*
On the Causes of the Variations to be Investigated   *175*
The Quantity of Reducing Phlogiston Sought by Experiments in the Humid Way   *178*
Corollaries of the Preceding Experiments   *188*
The Quantity of Reducing Phlogiston Sought by Experiments in the Dry Way   *192*
Corollaries of the Preceding Dry Experiments   *206*
The Quantity of Caloric Matter in Iron   *215*
Foreign Substances That are Present in Iron   *222*
The Proximate Principles of Iron   *234*
On the Calx of Iron   *244*
On Magnetism   *250*

CHAPTER NINE

Louis Bernard Guyton de Morveau

On the Nature of Steel and Its Proximate Principles   257–274

Extract from "Acier," *Encyclopédie Méthodique. Chymie, Pharmacie et Métallurgie* (Paris, 1786), Vol. I, pp. 447–51. Translated by P. Boucher and C. S. Smith.

CHAPTER TEN

Charles Auguste Vandermonde, Claude Louis Berthollet, and Gaspard Monge

On the Different Metallic States of Iron   275–348

"Mémoire sur le fer consideré dans ses différens états métalliques," from the *Mémoires de l'Académie Royale des Sciences*, 1786, pp. 132–200. Translated by Anne S. Denman and C. S. Smith.

    The Smelting of the Ore   *285*
    The Fining of Cast Iron   *288*
    Cementation of Iron   *291*
    Extract from the Researches of M. Réaumur   *297*
    Extract from the Researches of M. Bergman   *300*
    Account of Our Researches   *305*
    Explanation of the Processes That Are Followed in the Forges to Make Iron Pass through Its Different Metallic States   *328*
      On the smelting of ore   *328*
      On the fining of cast iron   *332*
      On the cementation of soft iron   *338*
    On Charcoal, Considered in Its State of Combination with Iron, and on Leaving That Combination   *340*
    Recapitulation   *347*

Index   349

Plates

Plate I.	*Von Stahel und Eysen.* Title page of first edition, Nuremberg, 1532.	2
Plate II.	Methods of testing steel bars. (Réaumur, 1722)	101
Plate III.	The fracture of a steel wedge quenched with a temperature gradient. (Réaumur, 1722)	103
Plate IV.	Diagrams of the internal structure of steel, and a machine for measuring "body." (Réaumur, 1722)	105
Plate V.	Crystals and blowholes in cast iron, with models of crystal structure. (Grignon, 1775)	157
Plate VI.	Dendritic crystals in cast iron. (Grignon, 1775)	159
Plate VII.	Grains and crystals in overheated iron. (Grignon, 1775)	161
Plate VIII.	Crystallizations in iron and iron slag. (Grignon, 1775)	163
Plate IX.	*Dissertatio Chemica de Analysi Ferri* by Torbern Bergman. Title page of first edition, Uppsala, 1781.	166

CHAPTER ONE

On Steel and Iron

The anonymous booklet, *Von Stahel und Eysen*...(Nuremberg, 1532).
Translated by Anneliese G. Sisco.

Von Stahel vnd Eysen/Wie man dieselbigen künstlich weych vnd hert machen sol/Allen Waffenschmiden/ Goldschmiden/Gürtlern/Sigil vñ Stempffelschneidern/sampt allen andern kunstbaren werckleuten/so mit Stahel vnd Eysen/ire arbeyts vbung treyben/Eynem yeden nach gelegenheyt zů geprauchen/fast nützlich zů wissen.

Mit vil andern künstlin: Wie man Gold vnd Silber farben/auff ein yedes Metall/ mancherley weise machen sol. Darzů auch wie man in Stahel vnd Eysen oder auff Waffen etzen sol. Deßgleychen auch mancherley art/ warm vnd kalt Eysen vnd Messing ꝛc. zů löthen.

M.D.xxxij.

Editor's Note

Plate I shows the title page of what is probably the earliest edition of this work, printed in Nuremberg in 1532 by Künigund Hergotin as a pamphlet of eleven pages. It was a popular little book and was often reprinted, but all editions are now excessively rare as neither content nor form appealed to early bibliophiles. The copy of the first edition, now in the Herzog August Bibliothek at Wolfenbüttel, is unique.[a]

[a] *At least two other editions were printed in 1532, one probably in Erfurt by M. Sachs (a copy now in the Landesbibliothek in Gotha) and the other in association with two other little works in the* Drey schöner kunstreicher Büchlein... *printed by Michael Blum in Leipzig, 1532. Its companions in this volume dealt respectively with dyeing — the* Allerley Matkel, *a treatise on stain removal that had been issued as a separate booklet by Peter Jordan of Mainz the same year, reprinted many times in various combinations, finally to appear in an English translation by Sidney Edelstein (Technology and Culture, 1964, 5, 297–321) — and with painters' colors and inks — the* Artliche Kunst mancherley weyse Dinten, *which had previously appeared by itself as a separate pamphlet printed in Nuremberg by Simon Dunckel in 1531. The* Stahel und Eysen *was issued again as an individual pamphlet by Peter Jordan at Mainz in 1534 (copy in editor's collection) and by Jacob Cammerlander at Strasbourg in 1539. A separate Jordan edition printed in Mainz in 1532 is referred to by H. W. Williams (see later), but this cannot now be traced and was apparently a war casualty. All of these editions bear the name of the work on the title page and are essentially identical in text, though spelling differences abound. The contents of the* Stahel und Eysen *were also reprinted in various forms and languages in innumerable editions of the books of household and workshop recipes that are collectively referred to as the* Kunstbüchlein *and were published mainly in Germany in the sixteenth and seventeenth centuries.*

The Kunstbüchlein *are of great interest as records, if somewhat precarious ones, of early technological chemistry. Their complex bibliography has been examined in detail by James Ferguson in his essay* "Some Early Treatises of Technological Chemistry" *and its supplements, published in the* Proceedings of the Philosophical Society of Glasgow, *1888, 19, 126–59; 1894; 25, 224–35; 1911, 43,*

The importance of etching in the history of art led to a publication by Hermann W. Williams, Jr., "*A sixteenth-century German treatise Von Stahel und Eysen translated with explanatory notes*" (Technical Studies in the Field of Fine Arts, *1935*, 4, 63–92). This was based on the Strasbourg 1539 edition and contains very useful notes. Perhaps no new translation was needed, but a critical metallurgical examination of the treatise did seem to be needed, and this is the result.

It is difficult to know the extent to which these recipes in Stahel und Eysen represent actual practice at the time it was first printed. There would undoubtedly have existed a certain secrecy on the part of successful toolmakers, and the general public would be ready enough

---

232–58; and 1912, 44, 149–89, as well as in the book by Ernst Darmstaedter, Berg-, Probier- und Kunstbüchlein (Munich, 1926). Neither of these bibliographers knew of any of the individual issues of Stahel und Eysen *mentioned here, but from the discussion of the contents it is clear that the* Stahel und Eysen *was incorporated into* Kunstbüchlein *editions published in Leipzig, 1532; Augsburg, 1535, 1537, and 1538; and Frankfurt, 1535, to mention only the earliest sixteenth-century editions in German. Dutch versions appeared in 1549, 1581, and later. Rather similar recipes for the etching of iron and steel, but not on hardening, had been published in the earliest of all the printed* Kunstbüchlein, *a Dutch book,* T Bouck vā Wondre (*1513, reprinted with commentary by H. G. T. Frencken, Roermond, 1934*), *and in the German* Rechter Gebrauch d'Alchimei (*Frankfurt?, 1531*), *from which recipes were reprinted successively in most of the catchall editions.*

The largest and most famous book of secrets is the one assembled by the pseudonymous Italian Alessio Piemontese, whose Secreti *first appeared in Venice in 1555. There were many later editions in Italian, Latin, French, and English. The worthy Alexis copied out the contents of* Stahel und Eysen, *in somewhat rearranged form, together with many other recipes on topics of all kinds ranging from casting to cosmetics. The English translation of the third part of* The Secrets of Alexis *that was published in London in 1562 has the distinction of providing the first printed description of the hardening of steel in the English language, though the Elizabethan English is often far from crystal clear, and in any case the technology, transmitted from Germany via Italy, does not reflect contemporary British practice. Another twice-translated version of the recipes dealing with steel and iron appears in* A profitable boke declaring dyuers approoued remedies... *printed in London in 1583 and again in 1588 and later. This was "taken out of Dutche, and Englished by L. M.," i.e., Leonard Mascall, evidently from one of the main series of* Kunstbüchlein.

to believe in exotic practices behind the almost magic transformation of steel during hardening.

In reading these recipes, it should be borne in mind that the hardening of steel has not always consisted of two operations (quenching followed by tempering) but that the earlier process was a single-stage one, an interrupted quench so managed that the steel came out of the quenching bath at the desired final hardness. Much of the mystery regarding the snail juice and other animal and vegetable concoctions that are here recommended for quenching was perhaps associated with this, for pure water could cool steel too fast. Only two of the recipes given in Stahel und Eysen *clearly relate to the two-stage process, and even these seem to be aimed at rectifying a defective quench rather than constituting a normal hardening procedure. Temper colors are not mentioned as an index of the correct tempering temperature but rather the evaporation of blood or tallow. Two of the recipes for direct quenching, however, do refer to the color of the steel as it comes from the bath. Supposedly the steel shed its scale during contact with the water, and it may be that organic matter in the bath prevented too prompt reoxidation.*

The sections in the Stahel und Eysen *that deal with etching are more straightforward than those on hardening. The ingredients are identifiable in both the corroding liquid and the paste[b] and in the varnish or wax used as a stop-off. Even the gold-colored varnishes for iron and tin are straightforward compositions.*

The beginnings of etching are obscure. Corrosive salts and vegetable acids have a long history as pickling media for removing oxides from annealed metals. The texture of Damascus swords and of the pattern-welded swords of the Merovingians and Vikings would scarcely have been visible if they had not been etched. Nevertheless, the earliest

[b] The etching materials are mainly based on ferrous sulfate, sal ammoniac, and vinegar. They would work slowly, by corroding the iron rather than dissolving it. The Dutch work, T Bouck vā Wondre (*see previous footnote*), also mentions the use of nitric acid for etching iron as well as other metals. The main use of nitric acid (which had been known from about A.D. 1300) was in parting gold-silver alloys.

written reference to the use of an etching reagent on iron is in the eighth-century chemical manuscript at Lucca, which contains a single recipe for producing a rough coppery surface on iron as a preliminary to gilding it. This gilding technique is described in slightly more detail in the somewhat later Mappae Clavicula, *and it undoubtedly led directly to the production of decorative designs by deep localized chemical attack. The earliest surviving examples of such work are pieces of armor dating from the fifteenth century.*

By the time that the Stahel und Eysen *appeared, etching had become an important art and was used to decorate some of the most magnificent parade armor, especially in Germany. (See James G. Mann, "The Etched Decoration of Armour," Proc. British Academy, 1942, 28, 17–44.) Very shortly after A.D. 1500, the armorer began to make etched iron plates that could be inked and printed for the graphic artist. Two centuries later etching played an important role in revealing to scientists the presence of carbon in steel, and three centuries later it laid the structure of metals open to microscopic observation. (See C. S. Smith,* A History of Metallography.)

The Stahel und Eysen *is far less technical than the* Probierbüchlein *that as practical metallurgical literature had preceded it by several years. Its curious introduction disclaims any connection with alchemy, though it attributes the discovery of the processes to the alchemists. The descriptions would have been of little help to anyone who had served an apprenticeship in a metalworker's shop, yet the large number of editions of the pamphlet and of the broader collections of recipes into which its information was later incorporated leave no doubt that the work met a popular demand in its time. Even though it was soon to be supplemented by the far more comprehensive and carefully written metallurgical treatise of Biringuccio, it achieves some distinction as the first printed and therefore widely available information on the subject of the hardening of steel.*

On Steel and Iron

HOW THEY CAN BE SKILLFULLY SOFTENED AND HARDENED, WHICH ALL ARMORERS, GOLDSMITHS, SILVERSMITHS,° ENGRAVERS OF SEALS AND DIES, AND ALL OTHER SKILLED ARTISANS WHO USE STEEL AND IRON IN DOING THEIR WORK WILL FIND USEFUL, EACH ACCORDING TO HIS NEEDS.

TOGETHER WITH MANY OTHER RECIPES: HOW VARIOUS METALS CAN BE GIVEN THE COLOR OF GOLD OR SILVER IN SEVERAL DIFFERENT WAYS; HOW TO ETCH INTO STEEL AND IRON OR UPON ARMOR; HOW IRON, BRASS, ETC., CAN BE SOLDERED BY DIFFERENT METHODS, EITHER HOT OR COLD.

1532

[(folio Ai, verso)]

Introduction

SINCE a number of diverse books on alchemy have appeared in print, it seems to me unnecessary to write any more about metals except for calling attention to, and commenting on, some artful recipes that will be very useful to those who work with metals and will help them to increase their understanding and knowledge of them. This printing is in no way intended for the benefit of alchemists, for they are concerned with different arts. Although the lesser arts have all been invented originally by

---

° *Gürtlern. Though literally "girdlers," these artisans made all kinds of buckles, ornaments, and household equipment, usually of silver, though not always.*

alchemists, they are only the beginnings and are child's play compared with the practices of the alchemist's art. Therefore, whenever these recipes were brought to me and taught me by reliable folk, I did not wish to keep them for myself but wished to make them accessible to everyone to whose work they pertain and who may profit therefrom, in particular armorers,[d] locksmiths, engravers of seals, etc., together with all others who work iron and steel. If fault should be found with some of the recipes, pray do not reject the whole book on that account, but help to improve these recipes on the basis of your own practice and experience. Perhaps the fault lies in the user himself, because he did not follow the instructions correctly: all arts require practice and long experience, and their mastery is only gradually acquired.

First I Shall Tell How to Make the Electuaries
that are to be taken daily by those who work metals in the fire, in order to be protected against noxious, poisonous fumes

Take a handful of garlic and an equal amount of walnuts. Crush both together. Take honey, let it get thoroughly clarified in a pan, then add the mixture of garlic and nuts and let it boil until it thickens. Let it cool and add one ounce[e] of theriac and ginger, one-quarter ounce of cloves, and one-quarter ounce of nutmeg, all thoroughly crushed and mixed together. Then, when you wish to work something in the fire, [Aii] eat in the morning, on an empty stomach, a piece of this electuary as large as a hazelnut and you will be protected against all noxious fumes.

[d] Waffenschmid. *Although today* Waffe *means arms and/or armor, it is clearly used in this treatise also to mean tool. Perhaps "ironsmiths" would be more correct here. Later we translate* Waffe *as armor, weapon, or tool, depending on the context.*

[e] *In the German the weights are given throughout in* loths, *the* loth *being half an ounce, 1/32 of the local pound. In most cases the word is used loosely, simply to give ratio, not absolute weights.*

Those working with mercury and arsenic should especially take these electuaries; afterward they should soak a piece of absorbent cotton in vinegar, make little plugs of it, and stop up their ears and nose therewith in order to be safe from the fumes. Do not make light of this; these fumes are very dangerous and harmful.

First, How Iron is to be Hardened
and some of the hardness drawn[f] again

Take the stems and leaves of vervain, crush them, and press the juice through a cloth. Pour the juice into a glass vessel and lay it aside. When you wish to harden a piece of iron, add an equal amount of a man's urine and some of the juice obtained from the little worms known as cockchafer grubs. Do not let the iron become too hot but only moderately so; thrust it into the mixture as far as it is to be hardened. Let the heat dissipate by itself until the iron shows gold-colored flecks, then cool it completely in the aforesaid water. If it becomes very blue, it is still too soft.[g]

---

[f] *Entlassen. A word which literally means to dismiss, discharge, or release. Although the modern word for tempering, anlassen, is related to this, it seems better not to use the word "temper" here. Williams confuses the picture by translating* härten *throughout as "temper," which would be correct only in the old meaning of the word. The use of the word "draw" in current U.S. (though not British) terminology is close to the old German meaning, and we have accordingly adopted it. The 1583 English version usually says "soften."*

[g] *So that the flavor of the old English may be enjoyed, we give the text of this paragraph as it appears on page 68 in the 1583 edition:*

To make yron or steele hard. Take the iuyce of Varuen, cald in Latine varbena, and strayne it into a glasse, and ye wil quenche any yron, take thereof, and put to of men's pisse, and the distilde water of wormes, so mixe altogether, and quenche therein so farre as ye will haue it hard, but take heede it be not too harde, therefore take it forth soone after, and let it coole of it self, for when it is well seasoned ye shall see golden spottes on your yron.

Also the common hardning of yron or steele, is in cold water, & snow water, so when the edge shall seeme blue after his hardning, signifieth a good sign, and a right hardning.

You may also take human excrement [*i.e.*, *slops*] after it has been distilled a second time and quench in that.

Or take red land snails, distill water therefrom, and then quench in this water.

Item: Old charred leather and half as much salt.

How to draw the hardness of iron

Let the human blood stand until water forms on top. Strain off this water and keep it. Then hold the hardened tools over a fire until they have become hot and subsequently brush them with a feather soaked in this water; they will devour the water and become soft.

Another way of drawing the hardness

Take clarified honey, fresh urine of a he-goat, alum, borax, olive oil, and salt; mix everything well together and quench therein. [(Aii, verso)]

Another recipe — When a piece has become too hard and you wish partly to draw the hardness

Take the piece that is too hard and carefully hold it over glowing charcoal until it has become hot. Rub it with tallow and let the tallow dry away on it over the heat; it will then acquire the correct hardness.

Another recipe for softening iron

Scrape horn onto a piece of leather; add sal ammoniac and urine. Wrap the iron in the leather, let the leather burn away on it, and the iron will become soft.

Another recipe for making [*iron*] soft and tough

Take camomile flowers, one part geranium, and one part vervain; put everything in a pot with hot water, cover the pot well so that no fumes can escape, boil well, and then quench therein.

Here Follow Recipes for Hardening Steel

The first and ordinary way of hardening steel is to use cold water. When the cutting edge is blue it has the correct hardness.

How to harden steel and make good cutting edges

Take the leaves and the root of the plant called oxtongue, boil them in water, and quench therein.

Another good method of hardening

Take dragonwort with its leaves and an equal amount of vervain; boil everything in clean water and then let the liquid get clear and cold. Throw [*the steel*] therein, and it will become good and hard.

You may also harden with mustard ground up in good vinegar.

Item: Take juice obtained from cockchafer grubs and juice obtained from stonewort and quench therein.

Item: Take human hair, boil it in water until it turns blood red and then quench therein.

Item: Take equal parts of radish juice, celery juice, and resin and quench therein.

Item: Take varnish, dragon's blood, horn scrapings, half as much [Aiii] salt, juice made from earthworms, radish juice, tallow, and vervain and quench therein.

It is also very advantageous in hardening if a piece that is to be hardened is first thoroughly cleaned and well polished.

So that the tools or whatever you wish to harden shall not split or crack in hardening

Take tallow, heat it, and pour it into a vessel that contains cold water. When it stands a finger thick upon the water, gently thrust whatever you wish to harden through the tallow so that it is hardened first in the tallow and then in the water.

How to harden tools

Take tadpoles, and the cockchafer grubs that are found when fields are plowed or harrowed; place a handful of each separately in a glass jar, salt well. They will turn into water; quench therein.

How to harden files

Harden files in linseed oil, or horn, or the blood of a he-goat.

How to harden [*file-*]cutting hammers, files, and other cutting tools

Crush up some radish, horse-radish, earthworms, cockchafer grubs, he-goat's blood, mixing everything together; sharpen the tool and harden it therein.

Item: A pick, a piece of armor,[h] any piece of steel, knives, or whatever you wish, may be quenched in beet juice. It gives a good hardening.

A hardening that cuts through everything

Distill snails, including their little spiral houses, and quench in the resulting water. Pieces that you wish to have very hard, first coat over well with ground sand and sulphur and then quench in this water.

How to harden augers, bits, and other tools

Stir well together equal amounts of a man's cold urine, vervain juice, and juice made of cockchafer grubs and quench therein as far as you wish the tool to harden. Let it cool by itself until little gold-colored flecks appear and then quench it completely in the aforesaid water. [(Aiii, verso)]

[h] Panzer. *The 1583 English reads "Mayles."*

Here Follows How Steel is to be Made Soft
so that it can be cut

Make a lye from equal amounts of willow ashes and unslaked lime, letting them stand to react for two hours. Immerse the steel in this lye for fourteen days. If you then wish to have it again as hard as it was before, place it in cold water.

Another recipe

Take equal parts of sal ammoniac and unslaked lime and a somewhat larger amount of Venetian soap than of these two. Work them up thoroughly together. After packing the steel in this mixture and moistening it with vinegar, place it in the fire, leave it there for three or four hours at most, then cool very gradually. This applies to small pieces; large ones you should treat as follows:

Take cow dung, egg white, and clay moistened with vinegar, and lay [*the steel*] in as before.

Another

Wrap up equal parts of salt and tartar inside a piece of clay, place the steel therein, then leave it in the fire for two hours and afterward allow it to cool down by itself.

Another

Take equal parts of soap and unslaked lime and a scant part of sal ammoniac; make a paste of this mixture and coat the steel therewith. Then lay some clay around it and let it be thoroughly annealed.

The Following Deals with Soldering
First: How iron is soldered cold

Take one ounce each of sal ammoniac, common salt, calcined tartar, and bell metal,[1] and three ounces of antimony [*sulphide*],

---

[1] *Unless this bell metal were high in tin, it would be very difficult to crush to powder. This composition seems to have been intended to join parts together by the formation of a compact corrosion product.*

everything thoroughly crushed and sifted, and tie all of it together in a linen cloth. Carefully coat the cloth all around with a finger's thickness of a well-prepared lute and let it get thoroughly dry. Then place it in an earthenware pot, cover this pot with a second, inverted, one and place it thus into a gently burning charcoal fire. Let it heat up gradually; then increase the fire until the ball is thoroughly red hot. [(Aiv)] Everything will then be melted together. Let it cool, break it open, and crush and grind the contents well so as to obtain a fine powder.

Then, if you wish to solder something, tack the pieces onto a board, bringing the joint together as accurately as you can. First, however, place a piece of paper underneath. Put a little of the aforesaid powder between and upon the joint; then build around it a little box of clay, which is open on top. Take borax, place it in warm wine until it dissolves; then take a feather and with it brush some of the wine on the powder. It will start to boil, and when it has stopped boiling, the piece will be repaired. Any material remaining on the joint must be ground off; it cannot be filed.

How to solder hot

Take gum water and crushed-up chalk, make a paste thereof and with it coat the joint. Then scrape off the paste where you wish to solder, but leave it on the surrounding area. Next coat the joint where you wish to solder with soap [*and apply some solder*]. Then hold a coal against it, and it will quickly melt. Afterward wash the paste off the soldered work.

Solder for copper

Half an ounce of copper to three-quarters of an ounce of white arsenic. Melt the copper; divide the arsenic in two parts, throw one part into the copper, stir, throw in the other part. Pour the mixture out onto a stone and hammer it thin.

Brass for soldering

File the brass quite small; add borax scrapings to it.

How to solder iron

File down the joint in the iron until the ends fit satisfactorily; then put it in a fire and throw Venetian glass on it. It will be soldered.[j]

How to make a powder that renders any metal easy-melting and malleable

Take four parts of antimony [*sulphide*] and one part each of sandiver and salt; work all this into a good powder. Use three parts of the powder to one part of metal and melt.

Another one for ore

Take prefused salt, tartar, saltpeter, sandiver, grapevine [(Aiv verso)] ashes or ashes of wine lees, and unslaked lime. Powder this and throw it on to the ore.

Another one

Take one ounce of unslaked lime, one and a half ounces of grapevine ashes or ashes of wine lees, two ounces of willow ashes, three ounces of beechwood ashes; mix all this well, pour water on it, and let it stand for two weeks until a lye has formed; strain it off. Let the ore become red hot and quench it therein; then crush it into small pieces, wash it, let it dry, and smelt.

Another one

Take filings, sift them; also take sulphur and lead and grind them into a powder; also take litharge, saltpeter, salt, sandiver,

---

[j] *Either the addition of solder is required, or this is a recipe for hammer-welding, the glass serving as a flux. (Solder was also omitted in the third recipe preceding, where melting was specifically referred to.) The 1583 English translation interpolates* "and so it shall sowder of himself, and bee very strong withall." *The word* löthen, *which we have always translated "to solder," can also mean to join, in general.*

and willow ashes, all of them together made into a good powder, and throw it on the ore.

How to Etch into Steel and Iron,
or upon armor[k]

Take one part of crushed lindenwood charcoal, two parts of vitriol, two parts of sal ammoniac and grind all this up with vinegar until it resembles a thick mash. Any inscription or design that is to be etched on a workpiece must first be written or drawn thereon with red lead tempered with linseed oil. Let it dry; then coat your work with a layer of the paste about as thick as a little finger and keep in mind that the hotter it is, the more quickly it will etch; be careful, however, not to burn the work. After the powder has thoroughly dried, remove it and wipe off the design.

Or: Take two parts of verdigris and one part of common salt, grind both in a mortar, add strong vinegar, and then proceed as before.

Or: Take vitriol, alum, salt, white vitriol, vinegar, and lindenwood charcoal, and proceed as before.

Or: Take two parts of vitriol and one-third part of sal ammoniac and grind both on a stone slab with urine. Use this as described above, except that the mixture must be used cold and the work must then be kept in a cellar for four or five hours.

Another method of etching, in which waters are used [Bi]

Take equal parts of verdigris, mercury sublimate, vitriol, and alum, all crushed fine. Place everything in a glass jar and let it stand for half a day, stirring frequently. Then take wax, or massicot mixed with linseed oil, or a mixture of red lead and linseed oil, and with it write on your workpiece any inscription that is to be etched thereon; then apply the water with a brush. Let it stand for half a

---

[k] *The distinction between "on" or "in" relates to the positive or negative character of the design or inscription.*

day. If the etching is to be very deep, let it stand correspondingly longer. If you wish to etch a depressed inscription or design, coat the iron or steel with a very thin layer of wax and write the inscription with a stylus into the wax through to the metal; then brush on the water, which will eat into it. Or lay mercury sublimate on the inscription that you have scratched in with the stylus, pour vinegar on it, and let it stand for half an hour.

Another, stronger one

One half ounce of verdigris, one-quarter ounce each of feather alum, sal ammoniac, tartar, vitriol, and common salt, all crushed fine and mixed together. Pour strong vinegar on this mixture and let it stand one hour. If your workpiece is to have a raised inscription, inscribe it with linseed oil and massicot and let it dry. Heat the above water in a glazed vessel, keep it on the fire, and hold the piece of steel or iron over the vessel. Pour some of the hot water over it with a spoon so that the water runs back into the vessel. Do this for a quarter of an hour and then scour it with ashes or with unslaked lime. See to it that the work be thoroughly coated with [*oil and*] massicot where it is [*not*] to be etched.

How to Give Different Metals the Color of Silver or Gold

First: A ground for gold or silver on iron, bells, stone, etc., that cannot be removed with water   [(Bi, verso)]

Take one part of ocher, one-third part of red lead, one-fourth part of Armenian bole, and an equal amount of spirits of wine; grind all this together with linseed oil and also grind in a piece of white vitriol as large as a hazelnut. Finally mix in three or four small drops of varnish. If, then, the paint is too thick, add more linseed oil. Subsequently transfer it from the stone [*grinding slab*] into a linen cloth and squeeze it out into a clean container. It should be as thick as honey. Spread it on to whatever you wish, let it dry, and apply the gold or silver leaf over it.

A gold paint for silver, tin, copper, etc.

Take a small glazed pot and place it in three ounces of linseed oil; then take half an ounce of mastic, half an ounce of Socotrine hepatic aloes,[1] both finely powdered, and mix them with the oil. Cover the pot with another identical one, turned upside down, which should have a hole in its bottom. Seal the joint well with good lute. Through the hole on top insert a small piece of wood, which should be broad at its lower end and may be used for stirring. Let it boil like painter's varnish. Whatever is to be gilded with this paint must first be thoroughly polished; then spread on the paint, let it dry in the sun, and, if the application was too thin, spread on more until you are satisfied.

Another one

Take varnish [*sic*], amber, and alum, both ground fine; then add varnish and linseed oil; boil all of it together in a glazed pot, on a charcoal fire, so that it melts and becomes thoroughly mixed together. Try it on a knife. If it is too thick, add linseed oil; but if it is too thin, add alum.

Another one

Take half an ounce of Socotrine hepatic aloes and one of amber, reduce both to a fine powder; put the powder in a glazed pot and place it on glowing coals, but do not let it get too hot at the start. When the content has become completely liquid, pour boiling oil on it, stir well with a piece of wood, let it cool, and strain it through a cloth. [Bii]

---

[1] Aloepaticum citrinum. *This interpretation follows Williams, who suggested that* Aloe hepaticum sicotrinum *was meant. Socotra (in the Gulf of Aden) was a well-known source of aloes.*

Another one

Melt on a charcoal fire half an ounce of Armenian bole and half an ounce of white gum; add one ounce of linseed oil; when it spins a thread, it is ready.

Gold color for tin

Take as much linseed oil as you wish. After it has been on the fire until it is clarified, add well-ground amber and hepatic aloes in equal amounts. Mix both well with the oil, on the fire, until the mixture thickens; then take it off and bury the pot, well covered, in the earth for three days. Anything of tin that you smear with this will acquire a golden color.

How to silver copper

Take tartar, alum, and salt; grind everything on a stone slab, add a leaf or two of silver, grind this in also, and then put it all in a glazed pot. Pour on water and toss the copper in. Scratch it with a brush, and you will see when it has had enough.

How to gild steel and iron

Take one part of tartar, half as much sal ammoniac, an equal amount of verdigris, and a little salt. Boil everything in white wine. Apply it to polished steel or iron with a brush and let it dry. Then gild with ground-up gold.

THE END

Printed at Nuremberg by Künigund Hergotin

CHAPTER TWO

Concerning the Method of Making Steel

Vannoccio Biringuccio

Extract from *De la Pirotechnia* (Venice, 1540). Translated...by C. S. Smith and M. T. Gnudi (New York: © American Institute of Mining, Metallurgical and Petroleum Engineers, 1942), pp. 67–70.

Editor's Note

This section on the making of steel is taken from the book De la Pirotechnia *by the Italian foundryman and metallurgist, Vannoccio Biringuccio. First printed in Venice in 1540, it is the earliest comprehensive book on metallurgy and a rich source on technology in general. It is thoroughly practical in outlook, but as behooved an educated man of his time, Biringuccio introduced many of his topics with a brief discussion of the Aristotelian elemental mixtures combined in the materials involved. His opening paragraph on steel repeats Aristotle's suggestion that steel is a pure form of iron, but it is followed by a description of the Brescian process of making steel by dunking a thirty- to forty-pound bloom of wrought iron into a bath of molten cast iron that is unmistakably practical. Biringuccio's description of steelmaking was copied almost verbatim by Georgius Agricola in his* De Re Metallica (1556), *with an added illustration of the hearth and forge.*

It should, perhaps, be mentioned that this steelmaking process was not widely used outside Italy. The production of cast iron in the blast furnace was only a little more than a century old when Biringuccio was writing, and much iron was still being made by direct reduction in a smelting hearth. Steel was generally made in similar hearths by holding the reduced sponge iron in prolonged contact with the charcoal fuel, but it required very skillful operation, and in many iron-smelting areas little steel was intentionally made. The process that Biringuccio describes may have been derived from the East, for something like it was done in the sixth century A.D. in China, where cast iron had been appreciated from the time of the first use of iron itself.[a]

Elsewhere in his book (Book IX, Chapter 6), Biringuccio discusses the many-sided work of the ironsmith and gives a number of "secrets" that seem to have been extracted from a manuscript collection not unlike the Stahel und Eysen, *though more diverse in content and obviously quite independent. The most complete of these recipes reads as follows:*

Other secrets are the various temperings with water, herb juices, or oils, as well as the tempering of files. In these things, as well as in common water, it is necessary to understand well the colors that are shown and thrown off on cooling. It is necessary to know how to provide that they acquire these colors well in cooling, according to the work and also the fineness of the steel. Because the first color that is shown by steel when it is quenched while fiery is white, it is called silver; the second which is yellow like gold they call gold; the third which is bluish and purple they call violet; the fourth is ashen gray. You quench them at the proper stage of these colors as you wish them more or less hard in temper. If you wish it very hard, heat your iron well and quench it rapidly in the tempering baths that you have prepared or in clear cold water.

Biringuccio soldered iron with a base silver alloy and a borax or glass flux; softened it by coating it with a composition of oil, glass, alkali, and dung and putting it in a fire which was allowed to go out; he etched it with "water" made of sal ammoniac, corrosive sublimate, verdigris, and nut gall in vinegar. He describes gilding and damascening, and concludes:

Thus, in conclusion, it seems to me an art comprising great knowledge, for I know of no art or activity whatever, excluding the sciences

---

[a] See J. W. Needham, Development of Iron and Steel Technology in China (London, 1958), and the discussion by J. R. Spencer and others of the first European description of a blast furnace, in Technology and Culture, 1963, 4, 201–2, and 1964, 5, 386–405.

and painting, that does not need this as its principal member. Therefore, in my opinion, if it were not for the nobility of the material, I would say that the smith working in iron should justly take precedence over the goldsmith because of the great benefit that he brings.

Concerning the Method of Making Steel

ALTHOUGH it might seem more fitting to discuss this subject in the Ninth Book in connection with the smelting of iron where I had thought to treat of it in detail, this process of making steel appears to be almost a branch of the above chapter on iron itself; hence I did not wish to separate these two so far that they might seem to be [18v][b] different things. Therefore I wish to write of it here and to tell you that steel is nothing other than iron, well purified by means of art and given a more perfect elemental mixture and quality by the great decoction of the fire than it had before. By the attraction of some suitable substances in the things that are added to it, its natural dryness is mollified by a certain amount of moisture and it becomes whiter and more dense so that it seems almost to have been removed from its original nature. Finally, when its pores have been well dilated and softened by the strong fire and the heat has been driven out of them by the violence of the coldness of the water, these pores shrink and it is converted into a hard material which, because of the hardness, is brittle.

This steel[c] can be made from any kind of iron ore or prepared iron. It is indeed true that it is better when made from one kind than from another and with one kind of charcoal than with another;

[b] *Numbers in brackets refer to pages in the first edition, Venice, 1540.*

[c] *The process consists of immersing masses of wrought iron in a bath of molten cast iron. Although Biringuccio says that both the bloom and the bath are of the same iron, that for the bath is undoubtedly carburized by the excess of charcoal mixed with it for melting.*

Agricola (De re metallica [Basle, 1556], p. 342) *copies this description almost word for word and adds an illustration showing steel bars being hammered and quenched. Unfortunately the hearth of the furnace is not shown and it is possible that it is just a forging operation that is depicted, not the steel-making process itself.*

it is also made better according to the understanding of the masters. Yet the best iron to use for making good steel is that which by nature is free from corruption of other metals and hence is more disposed to melt and has a somewhat greater hardness than the other. Crushed marble or other rocks readily fusible in smelting are placed with this iron; these purify the iron and almost have the power to take from it its ferruginous nature, to close its porosity, and to make it dense and without laminations.

Now in short, when the masters wish to do this work they take iron that has been passed through the furnace or obtained in some other way, and break into little pieces the quantity that they wish to convert into steel. Then they place in front of the tuyère of the forge a round receptacle, half a *braccio* or more in diameter, made of one-third of clay and two-thirds of charcoal dust, well pounded together with a sledge hammer, well mixed, and then moistened with as much water as will make the mass hold together when it is pressed in the hand. And when this receptacle has been made like a cupeling hearth but deeper, the tuyère is attached to the middle so that its nose is somewhat inclined downwards in order that the blast may strike in the middle of the receptacle. Then all the empty space is filled with charcoal and around it is made a circle of stones or other soft rocks which hold up the broken iron and the charcoal that is also placed on top; thus it is covered and a heap of charcoal made. Then when the masters see that all is afire and well heated, especially the receptacle, they begin to work the bellows more and to add some of that iron in small pieces mixed with saline marble,[d] crushed slag, or other fusible and nonearthy stones. Melting it with such a composition they fill up the receptacle little by little as far as desired.

Having previously made under the forge hammer[e] three or four blooms weighing thirty to forty pounds each of the same iron, they put these while hot into this bath of molten iron. This bath

[d] marmo saligno.
[e] *forge hammer* — maglio.

is called "the art of iron"[f] by the masters of this art. They keep it in this melted material with a hot fire for four or six hours, often stirring it up with a stick as cooks stir food. [19] Thus they keep it and turn it again and again so that all that solid iron may take into its pores those subtle substances that are found in the melted iron, by whose virtue the coarse substances that are in the bloom are consumed and expanded, and all of them become soft and pasty. When the masters observe this they judge that the subtle virtue has penetrated fully within; and they make sure of it by testing, taking out one of the masses and bringing it under a forge hammer to beat it out, and then, throwing it into the water while it is as hot as possible, they temper[g] it; and when it has been tempered they break it and look to see whether every little part has changed its nature and is entirely free inside from every layer of iron. When they find that it has arrived at the desired point of perfection they take out the lumps with a large pair of tongs or by the ends left on them and they cut each one in six or eight small pieces. Then they return them to the same bath to heat again and they add some more crushed marble and iron for melting in order to refresh and enlarge the bath and also to replace what the fire has consumed. Furthermore, by dipping that which is to become steel in this bath, it is better refined. Thus at last, when these pieces are very hot, they are taken out piece by piece with a pair of tongs, carried to be drawn out under the forge hammer, and made into bars as you see. After this, while they are still very hot and almost of a white color because of the heat, in order that the heat may be quickly quenched they are suddenly thrown into a current of water that is as cold as possible, of which a reservoir has been made.

In this way the steel takes on that hardness which is commonly called temper; and thus it is transformed into a material that

[f] larte di ferro. Mieli suggests latte di ferro, "*milk of iron.*"

[g] *The word "tempering" refers to the quenching operation, and not, as now, to reheating after a drastic quench. It produced a finely tempered product, however, for the quenching operation was controlled by intermittent, slow, or partial immersion to give the desired hardness directly.*

scarcely resembles what it was before it was tempered. For then it resembled only a lump of lead or wax, and in this way it is made so hard that it surpasses almost every other hard thing. It also becomes very white, much more so than is the nature of the iron in it; indeed it is almost like silver. The kind that has a white, very fine, and fixed grain is the best. The kinds I have heard of that are highly praised are that of Flanders and, in Italy, that of Valcamonica in the Brescian district. Outside Christendom the Damascan is praised and the Chormanian, the Azziminan, and that of the Agiambans. I do not know how those people obtain it or whether they make it, although I was told that they have no other steel than ours. They say that they file it, knead it with a certain meal, make little cakes of it, and feed these to geese. They collect the dung of these geese when they wish, shrink it with fire, and convert it into steel. I do not much believe this, but I think that whatever they do is by virtue of the tempering, if not by virtue of the iron itself.

CHAPTER THREE

Of Tempering Steel

Giovanni Battista della Porta

Extract from *Magiae Naturalis Libri Viginti* (Naples, 1589). Reproduced from the anonymous English translation (London, 1658), pp. 305–12.

Editor's Note

Most delightful and browsable of scientific books, yet most valuable by virtue of its content and historical significance, *Natural Magick* was a best-seller from its first appearance. The very title gives an indication of the book's peculiar charm and durable popularity. It gives also a warning to the modern reader that he must have his full historical senses about him if he is not to be misled by words and thoughts that have changed their fabric in three or four centuries.

With these words Derek Price appropriately prefaces his modern reprinting (New York, 1957) of the seventeenth-century English translation of Giambattista della Porta's Magia Naturalis.[a] *Porta (ca. 1535–1615) clearly distinguishes between sorcery and "natural" magic, which is constructive. His book is an amazing collection of household recipes, craftsmen's secrets, simply curious observations, and portentous experiments. There are hints of later optical instruments and a discourse "On the Causes of Wonderful Things." Alchemy becomes natural, almost modern, in the chapters telling how to change the colors of metals either by superficial treatment or by alloying. Other chapters give recipes for colored glass for imitation gems and describe the making of tinted metal foils to place behind cut stones. Magnetism, distillation, and invisible writing all caught Porta's interest. In other works he discussed physiognomy and laid some basis for scientific meteorology.*

[a] *The first edition of the complete work was in Latin*, Magiae Naturalis Libri Viginti...(*Naples, 1589*). *It was frequently reprinted and translated into Italian, French, German, and English. The English translation by an anonymous translator was printed in London in 1658, reprinted in 1669. Our excerpt is photographically reproduced from the first English edition.*

He mentions the compression of air by means of a jet of water, used as a substitute for bellows in providing blast for a furnace, and the cooling of a room by the (adiabatic) expansion of compressed air. The material on the tempering of steel that is reproduced below — the Thirteenth Book of the Magia Naturalis *occupying pages 219–25 of the 1589 edition — is sandwiched between one on artificial fires (incendiary mixtures) and "Some Choice Things on the Art of Cookery."*

Though there are parts of Porta's book that seem to reflect the earlier books of secrets, most of it is clearly original. Certainly there is no antecedent for the chapters on the tempering of steel, and the almost magical changes in properties produced by quenching steel were clearly of the type to arouse the author's especial interest. This section is lacking in the earlier (1558) edition of the Magia Naturalis *in four books and first appeared in the twenty-book 1589 edition.*

*Of particular interest to a metallurgist are Chapter 3, discussing the different manners of hardening iron (i.e., steel) for different purposes, and Chapter 4 on the cementation (case hardening) of iron files and armor. Porta clearly describes quenching and subsequent tempering — "The workmen call it a return" (*revenire *in Italian). He is indeed the first author who clearly prefers a two-stage process of quenching, cleaning, and tempering, to that of direct quenching. However, he also provides in Chapter 6 an excellent description of the interrupted quench, with emphasis on the necessity for "clear and purified juices" to enable the colors to be discerned. It will also be noted that Porta, like Biringuccio before him, was aware of the value of steel from the Orient. He tells how to renew the "wave marks" on a Damascus sword and also how to fake the texture by etching after appropriate stopping off.*

THE THIRTEENTH BOOK OF Natural Magick:

Of tempering Steel.

THE PROEME.

I Have taught you concering monstrous Fires; and before I part from them, I shall treat of Iron Mines; for Iron is wrought by Fire: not that I intend to handle the Art of it; but onely to set down some of the choicest Secrets that are no less necessary for the use of men, in those things I have spoken of already, besides the things I spake of in my Chymical works. Of Iron there are made the best and the worst Instruments for the life of man, saith Pliny. For we use it for works of Husbandry and building of Houses; and we use it for Wars and Slaughters: not onely hard by; but to shoot with Arrows, and Darts, and Bullets, far off. For, that man might die the sooner, he hath made it swift, and hath put wings to Iron. I shall teach you the divers tempers of Iron, and how to make it soft and hard, that it shall not onely cut Iron and other the hardest substances, but shall engrave the hardest Porphyr and Marble Stones. In brief, the force of Iron conquers all things.

CHAP. I.
That Iron by mixture may be made harder.

T is apparent by most famous and well-known Experience, that Iron will grow more hard by being tempered, and be made soft also. And when I had sought a long time whether it would grow soft or hard by hot, cold, moist or dry things; I found that hot things would make it hard and soft, and so would cold and all the other qualities: wherefore something else must be thought on to hunt out the causes. I found that it will grow hard by its contraries, and soft by things that are friendly to it; and so I came to Sympathy and Antipathy. The Ancients thought it was done by some Superstitious Worship, and that there was a Chain of Iron by the River Euphrates, that was called Zeugma, wherewith *Alexander the Great* had there bound the Bridge; and that the links of it that were new made, were grown rusty, the other links not being so. *Pliny* and others think, That this proceeded from some different qualities; it may be some juices or Minerals might run underneath, that left some qualities, whereby Iron might be made hard or soft. He saith, But the chief difference is in the water that it is oft plunged into when it is red hot. The pre-eminence of Iron that is so profitable, hath made some places famous here and there; as Bilbilis and Turassio in Spain, Comum in Italy: yet are there no Iron Mynes there. But of all the kindes, the Seric Iron bears the Garland; in the next place, the Parthian: nor are there any other kindes of Iron tempered of pure Steel: for the rest are mingled. *Justine* the Historian reports, That in Gallicia of Spain, the chiefest matter for Iron is found; but the water there is more forcible then the Iron: for the tempering with that, makes the Iron more sharp; and there is no weapon approved amongst them, that is not made of the River Bilbilis, or tempered with the water of Chalybes. And hence are those people that live neer this River called Chalybes; and they are held to have the best Iron. Yet *Strabo* saith, That the Chalybes were people in Pontus neer the River Thermodon. *Virgil* speaks,

And the naked Calybes Iron.

Then, as *Pliny* saith, It is commonly made soft with Oyl, and hardened by Water. It is a custome to quench thin Bars of Iron in Oyl, that they may not grow brittle by being quenched in Water. Nothing hath put me forward more to seek higher matters, then this certain Experiment, That Iron may be made so weak and soft by Oyl, that it may be wrested and broken with ones hands: and by Water it may be made so hard and stubborn, that it will cut Iron like Lead.

Chap. II.
How Iron will wax soft.

I Shall first say how Iron may grow soft, and become tractable; so that one may make steel like Iron, and Iron soft as Lead. That which is hard, grows soft by fat things, as I said; and without fat matter, by the fire onely, as *Pliny* affirms. Iron made red hot in the fire, unless you beat it hard, it corrupts: as if he should say, Steel grows soft of it self, if it be oft made red hot, and left to cool of it self in the fire: and so will Iron grow softer. I can do the same divers wayes.

That Iron may grow soft,

Anoynt Iron with Oyl, Wax, Asafœtida; and lute it over with straw and dung, and dry it: then let it for one night be made red hot in burning coals. When it grows cold of it self, you shall finde it soft and tractable. Or, take Brimstone three parts, four parts of Potters Earth powdered: mingle these with Oyl to make it soft. Then cover the Iron in this well, and dry it, and bury it in burning coals: and, as I said, you may use Tallow and Butter the same way. Iron wire red hot, if it cool alone, it will be so soft and ductible, that you may use them like Flax. There are also soft juices of Herbs, and fat, as Mallows, Bean-Pods, and such-like, that can soften Iron; but they must be hot when the Iron is quenched, and Juices, not distilled Waters: for Iron will grow hard in all cold waters, and in liquid Oyl.

Chap. III.
The temper of Iron must be used upon soft Irons.

I Have said how Iron may be made softer, now I will shew the tempering of it, how it may be made to cut sharper. For the temper of it is divers for divers uses. For Iron requires several tempers, if it be to cut Bread, or Wood, or Stone, or Iron, that is of divers liquors; and divers ways of firing it, and the time of querching it in these Liquors: for on these doth the business depend. When the Iron is sparkling red hot, that it can be no hotter, that it twinkles, they call it Silver; and then it must not be quenched, for it would be consumed. But if it be of a yellow or red colour, they call it Gold or Rose-colour: and then quenched in Liquors, it grows the harder: this colour requires them to quench it. But observe, That if all the Iron be tempered, the colour must be blew or Violet colour, as the edge of a Sword, Rasor or Lancet: for in these the temper will be lost if they are made hot again. Then you must observe the second colours; namely, when the Iron is quenched, and so plunged in, grows hard. The last is Ash colour: and after this if it be quenched, it will be the least of all made hard. For example:

The

The temper of a Knife to cut Bread.

I have seen many ingenious men that laboured for this temper, who, having Knives fit to cut all hard substances, yet they could scarce fall upon a temper to cut Bread for the Table. I fulfilled their desire with such a temper. Wherefore to cut Bread, let the Steel be softly tempered thus: Heat gently Steel, that when its broken seems to be made of very small grains; and let it be excellent well purged from Iron: then strike it with a Hammer to make a Knife of it: then work it with the File, and frame it like a Knife, and polish it with the Wheel: then put it in o the Fire, till it appear Violet-colour. Rub it over with Sope, that it may have a better colour from the Fire: then take it from the Fire, and anoynt the edge of it with a Linen-cloth dipt in Oyl of Olives, until it grow cold; so you shall soften the hardness of the Steel by the gentleness of the Oyl, and a moderate heat. Not much differs from this,

The temper of Iron for Wood.

Something harder temper is fit to cut wood; but it must be gentle also: therefore let your Iron come to the same Violet-colour, and then plunge it into waters: take it out; and when it appears Ash-colour, cast it into cold water. Nor is there much difference in

The temper for Instruments to let blood.

It is quenched in Oyl, and grows hard; because it is tender and subtile: for should it be quenched in water, it would be wrested and broken.

The temper of Iron for a Sythe.

After that the Iron is made into a Sythe, let it grow hot to the colour of Gold, and then quench it in Oyl, or smeer it with Tallow, because it is subtile Iron; and should it be quenched in waters, it would either crumble or be wrested.

Chap. IV.
How for all mixtures, Iron may be tempered most hard.

NOw I will shew some ways whereby Iron may be made extream hard: for that Iron that must be used for an Instrument to hammer, and polish, and fit other Iron, must be much harder then that.

The temper of Iron for Files.

It must be made of the best Steel, and excellently tempered, that it may polish, and fit o her Iron as it should be: Take Ox hoofs, and put them into an Oven to dry, that they may be powdered fine: mingle well one part of this with as much common Salt, beaten Glass, and Chimney-soot, and beat them together, and lay them up for your use in a wooden Vessel hanging in the smoak; for the Salt will melt with any moisture of the place or Air. The powder being prepared, make your Iron like to a file: then cut it chequerwise, and crosswayes, with a sharp edged tool: having made the Iron tender and soft, as I said, then make an Iron chest fit to lay up your files in, and put them into it, strewing on the powders by course, that they may be covered all over: then put on the cover, and lute well the chinks with clay and raw, that the smoak of the powder may not breath out; and then lay a heap of burning coals all over it, that it may be red-hot about an hour: when you think the powder to be burnt and consumed, take the chest out from the coals with Iron pinchers, and plunge the files into very cold water, and so they will become extream hard. This is the usual temper for files; for we fear not if the files should be wrested by cold waters. But I shall teach you to temper them excellently

Another way.

Take the pith out of Goats horns, and dry it, and powder it: then lay your files in a little Chest strewed over with this Powder, and do as you did before. Yet observe this, That two files supernumerary must be laid in, so that you may take them forth at pleasure: and when you think the Chest, covered with burning coals, hath taken in the force of the Powder, take out one of the supernumerary Files, and temper it, and break it; and if you finde it to be very finely grain'd within, and to be pure Steel, according to your desire, take the Chest from the fire, and temper them all the same way: or else, if it be not to your minde, let them stay in longer; and resting a little while, take out the other supernumerary File, and try it, till you have found it perfect. So we may

Temper Knives to be most hard.

Take a new Ox hoof, heat it, and strike it with a Hammer on the side; for the pith will come forth: dry it in an Oven; and, as I said, put it into a pot, alwayes putting in two supernumeraries, that may be taken forth, to try if they be come to be pure Steel; and doing the same as before, they will be most hard. I will shew

How an Habergeon or Coat of Arms is to be tempered.

Take soft Iron Armour of small price, and put it into a pot, strewing upon it the Powders abovesaid; cover it, and lute it over, that it have no vent, and make a good Fire about it: then at the time fit, take the Pot with iron pinchers; and striking the Pot with a Hammer, quench the whole Herness, red hot, in the foresaid water: for so it becomes most hard, that it will easily resist the strokes of Poniards. The quantity of the Powder is, that if the Harness be ten or twelve pounds weight, lay on two pounds and a half of Powder, that the Powder may stick all over: wet the Armour in water, and rowl it in the Powder, and lay it in the pot by courses. But, because it is most hard, lest the rings of a Coat of Male should be broken, and flie in pieces, there must be strength added to the hardness. Workmen call it a Return. Taking it out of the Water, shake it up and down in Vinegar, that it may be polished, and the colour be made perspicuous: then make red hot a plate of Iron, and lay part of the Coat of Male, or all of it upon the same: when it shews an Ash-colour, workmen call it Berotinum: cast it again into the water, and that hardness abated; and will it yield to the stroke more easily: so of a base Coat of Male, you shall have one that will resist all blows. By the mixture of Sharp things, iron is made hard and brittle; but unless strength be added, it will flie in pieces with every blow: therefore it is needful to learn perfectly how to add strength to it.

Chap. V.
Liquors that will temper Iron to be exceeding hard.

I Said that by Antipathy Iron is hardened, and softened by Sympathy: it delights in fat things, and the pores are opened by it, and it grows soft: but on the contrary, astringent things, and cold, that shut up the pores, by a contrary quality, make it extreme hard; they seem therefore to do it: yet we must not omit such things as do it by their property. If you would have

A Saw tempered to saw Iron,

Make your Saw of the best Steel, and arm it well that it be not wrested by extinguishing it. Then make a wooden Pipe as long as the Iron of the Saw, that may contain a liquor made of Water, Alom, and Piss; Plunge in the red hot Iron, and take it out, and observe the colours: when it comes to be violet, put all into the liquor, till it grow cold. Yet I will not conceal, that it may be done by a Brass wire bent like a bow, and with Powder of Emril and Oyl: for you shall cut Iron like Wood. Also, there are tempered

Fish-hooks to become extream hard.

The Hook serves for a part to catch Fish; for it must be small and strong: if it be great, the Fish will see it, and will not swallow it; if it be too small, it will break with great weight and motion; if it be soft, it will be made straight, and the Fish will get off.

off. Wherefore, that they may be strong, small, and not to be bended in the mouth; you shall thus temper them: Of Mowers Sythes make wire, or of the best Steel, and make Hooks thereof, small and fine: heat them not red-hot in the Fire; for that will devour them: but lay them on a plate of red hot Iron. When they grow red, cast them into the water: when they are cold, take them out and dry them. Then make the plate of Iron hot again, and lay on the Hooks the second time; and when an Ash colour, or that they commonly call Berotinus, appears, plunge them into the water again, that they may be strong: for else they would be brittle. So you may make

Culters extream hard.

Albertus, from whom others have it, saith, That Iron is made more strong, if it be tempered with juice of Radish, and Water of Earth-worms, three or four times: But I, when I had often tempered it with juice of Radish, and Horse-Radish, and Worms, I found it alwayes softer, till it became like Lead: and it was false, as the rest of his Receits are. But thus shall you make Steel extream hard, that with that onely, and no other mixture, you may make Culters very hard: Divide the Steel into very small pieces like Dice, and let them touch one the other, binding Iron wires over them, fastning all with an Iron wire: put them into the Fire till they grow red hot, and sparkle, at least fifteen times, and wrap them in these powders that are made of black Borax one part, Oyster shells, Cuttle-bones, of each two parts: then strike them with a Hammer, that they may all unite together, and make Culters, or Knives, or what you will: for they will be extream hard. For this is the most excellent sort of Steel, that onely tempered with waters, is made most hard. There is another, but not so good; and unless it be well tempered, it alwayes grows worse. It is this:

To temper a Graver to cut Marble.

Make your Graver of the best Steel, let it be red hot in the Fire, till it be red or Rose coloured; dip it into water, then take it away, and observe the second colour. When it is yellow as Gold, cast it into the water. So almost is

A Tool made to cut Iron.

When the same red Rose colour appears, plunge it into the water, or some sharp liquor that we shall shew; and you must observe the second yellow colour, or wheat colour, and then cast it into the water. These are the best

Tempers for Swords.

Swords must be tough, lest whilst we should make a thrust, they should break; also, they must have a sharp edge, that when we cut, they may cut off what we cut. The way is thus: Temper the body of it with Oyl and Butter, to make it tough; and temper the edge with sharp things, that they may be strong to cut: and this is done, either with wooden Pipes, or woollen Cloths, wet with Liquor: use it wittily and cunningly.

Chap. VI.
Of the temper of a Tool shall cut a Porphyr Marble Stone.

OUr Ancestors knew well to temper their Tools, wherewith they could easily cut a Porphyr Stone, as infinite Works testifie that were left to us: but the way was shewed by none, and is wholly concealed; which is a mighty disgrace to our times, when we neglect such rare and useful Inventions, and make no account of them. That we might be freed from this dishonour, with great care, and pains, and cost, I made trial of all things came to my hand, or I could think of, by divers wayes and experiments, that I might attain unto it: at last, by Gods great blessing, I found a far greater passage for to come to these things, and what exceeds this. And I will not be grieved to relate what I found out by chance, whilst I made trial of these things.

things. The business consisted in these difficulties. If the temper of the Graver was too strong and stubborn, with the vehement blow of the Hammer it flew in pieces: but if it was soft, it bowed, and would not touch the stone: wherefore it was to be most strong and tough, that it might neither yield to the stroke, nor flie asunder. Moreover, the juice or water the Iron must be tempered in, must be cleer and pure: for if it be troubled, the colours coming from heat could not be discerned: and so the time to plunge the Tools in would not be known, on which the whole Art depends. So then, cleer and purified juices will shew the time of the temper. The colours must be chiefly regarded: for they shew the time to plunge it in and take it out; and because that the Iron must be made most hard and tough, therefore the colour must be a middle colour between silver and gold: and when this colour is come, plunge the whole edge of the Tool into the liquor, and after a little time, take it out; and when it appears a Violet-colour, dip it into the liquor again, left the heat, yet remaining in the Tool, may again spoil the temper: yet this we must chiefly regard, that the liquors into which the Iron is plunged, be extream cold: for if they be hot, they will work the less: and you must never dip an Iron into water, that other Iron hath been dipt in before; for when it is grown hot, it will do nothing: but dip it into some other that is fresh and cold; and let this in the mean time, swim in some glazed Vessel of cold water, that it may soon grow cold, and you shall have it most cold for your work. Yet these are

The hardest tempers of Iron.

If you quench red-hot Iron in distilled Vinegar, it will grow hard. The same will happen, if you do it into distilled Urine, by reason of the Salt it contains in it. If you temper it with dew, that in the month of *May* is found on Vetches Leaves, it will grow most hard. For what is collected above them, is salt; as I taught elsewhere out of *Theophrastus*. Vinegar, in which Salt Ammoniac is dissolved, will make a most strong temper: but if you temper Iron with Salt of Urine and Salt-Peter dissolved in water, it will be very hard; or if you powder Salt-Peter and Salt Ammoniac, and shut them up in a Glass Vessel with a long neck, in dung, or moist places, till they resolve into water, and quench the red-hot Iron in the water, you shall do better. Also, Iron dipped into a liquor of quick Lime, and the Salt of Soda purified with a Spunge, will become extream hard. All these are excellent things, and will do the work: yet I shall shew you some that are far better.

To temper Iron to cut Porphyr Marble.

Take the fugitive servant, once received, and then exalted again, and shut it in a glazed Vessel, till it consume in Fire or water; so the Iron Tool will grow hard, that you may easily have your desire: but if it be too hard, that it be too brittle, add more liquor, or else more Metal: yet take care of this alone, whilst you have found the measure of your work: for the Iron will grow strong and tough. The same also will be happily performed by the foul moysture of the Serpent Python, and by the wasting thereof: for the salt gives force, and the fat toughness. And these are the best and choicest that I have tried in this kinde.

Chap. VII.
How to grave a Porphyr Marble without an Iron tool.

Some have attempted to do this without any Graver, but with strong and forcible water; and this Argument moved them to it: When they saw Vinegar and sharp juices to swell into bubbles, being cast upon Marble, and to corrode it, they supposed that if they should draw very strong sharp liquor from sharp and corroding things, they might do the same work without labour. At last, thus they did it: Take a little Mercury sublimate, and a little Salt Ammoniac, distil these as I shewed in Glass Stills: then take a little Verdigrease, Tin calcined, and of the fire-stone, powder all these with *Sal Gemma*, and common Salt, and Salt Ammoniac, and distil them, and pour
the

the distilled liquor again upon the Feces, and distil it again, and do it again the third time: then keep the liquor in a Vessel well stopt. When you go about your work, smeer the Porphyr Marble with Goats suet, onely touch not those parts you mean to have engraved: you must make a ledge about it, that when you pour on your water, it may not run off here and there; and the liquor poured on will eat most strongly: when it ceaseth to eat, cast it away, and pour on fresh; and do this so often, till you have graved it so much as you please, and you have done.

Chap. VIII.
How Iron may be made hot in the fire to be made tractable for works.

Many seek most diligently, how by a secret Art Iron may be so tempered, that it may neither break, nor be shot through with Guns. But these men do not take care of what they have before them, and seek for what they have not; for would they consider whilst the Iron heats, the thing they seek for so eargerly, is before their eyes. I say therefore, That the reason why Swords break and flie in pieces, and brests of Iron are shot through with Guns, is, because there are flaws in the Iron, and it cleaves in divers places, and the parts are ill united; and because these clefts are scarce visible: this is the cause that when they are bended or stricken they break: for if you mark well, whenever Knives or Swords break in pieces, you shall alwayes finde these craks and flames, and the solid parts are not broken; and being bended, resist. But when I sought for the cause of these flaws, I found at last, that in Smiths Shops, where Iron is made hot, they heap up coals over the Iron, and the refuse of coals; saying, The Iron will not heat so easily, if some rubbish of the coals and dust be not heaped over it: and with this trumpery-stuff, there are always mingled small stones, chalk, and other things gathered together in pieces; which, when they meet in the fire, they cause many knots outwardly, or cavities inwardly, and cracks, that the parts cannot well fasten together. Whence, though the business be trivial and of small regard, yet this is the cause of so great inconveniences that follow. Wherefore, to avoid this impediment, I thought on this course to be taken: I cast my coals into a wooden bowl full of water: for they will swim on the top, (but the filth and bricks will fall to the bottom) those that swim, I take out and dry them; and those I use for my works. What a blessing of God this profitable Invention is! for thus men make Swords, Knives, Bucklers, Coats of Male, and all sorts of Armour so perfect, that it were long and tedious to relate: for I have seen Iron brests, that scarce weighed above twelve pound, to be Musket-proof. And if we should add the temper to them, they would come to far greater effects.

Chap. IX.
How Damask Knives may be made.

Now whilst I set down these Operations very pleasant, namely, how Damask Knives may be made to recover their marks that are worn out, and how the same marks may be made upon other Knives. If then we would

Renew the waved marks of Damask Knives that are worn out,

polish a Poniard, Sword or Knife, very well with Powder of Emril and Oyl, and then cleanse it with Chalk, that no part may be dark, but that it may glister all over: then wet it all with juice of Lemmons mingled with Tanners water, that is made with Vitriol: for when it is dry, the marks will all be seen in their places, and wave as they did before. And if you will

Make marks with Damask Knives,

And that so acurately, that you can scarce know them from Damask Knives: Polish a Knife very well, as I said, and scowre it with Chalk: then stir with your hands,
Chalk

Chalk mingled with water; and touching it with your fingers, rub the edge of the Sword that was polished, and you shall make marks as you please: when you have done, dry them at the fire or Sun: then you must have a water ready wherein Vitriol is dissolved, and smeer that upon it: for when the Chalk is gone, it will dye it with a black colour. After a little stay, wet it in water, and wash it off: where the Chalk was, there will be no stain; and you will be glad to see the success. You may with Chalk make the waving Lines running up and down. If any one desires

To draw forth Damask Steel for work.

You may do it thus: for without Art it is not to be done. Too much heat makes it crumble, and cold is stubborn: but by Art, of broken Swords Knives may be made very handsomely; and Wheels and Tables, that Silver and Gold wire are drawn through, and made even by, to be used for weaving: Put it gently to the fire, that it may grow hot to a Golden colour, but put under the fire for ashes, Gip calcined, and wet with water: for without Gip, when you hammer it, it will swell into bubbles, and will flie and come to be dross and refuse.

Chap. X.
How polished Iron may be preserved from rust.

IT is so profitable to preserve Iron from rust, that many have laboured how to do it with ease. *Pliny* saith, That Iron is preserved from rust, by Ceruss, Gip, and liquid Pitch. But he shews not how Ceruss may be made: Yet those that know how to make Oyl of Ceruss without Vinegar, Iron being smeered therewith, is easily preserved from rust. Some anoynt the Iron with Deers suet, and so keep it free from rust; but I use the fat substance in the Hoofs of Oxen.

CHAPTER FOUR

The Selection of Iron and Steel, and the Manner of Hardening It by Quenching

Mathurin Jousse

Extract from *La Fidelle Ouverture de l'Art de Serrurier* (La Flèche, 1627). Reprinted with minor changes in footnotes from the English translation by C. S. Smith and A. G. Sisco in *Technology and Culture*, 1961, 2, pp. 131–45 (with permission of the Society for the History of Technology).

Editor's Note

The seventeenth century, which saw the birth of many concepts in the pure sciences that were later to influence metallurgy, saw remarkably little in the way of original additions to metallurgical literature. There were plenty of improvements in the scale and economy of production, but the book market was largely satisfied with reprints of the great sixteenth-century works — Biringuccio, Agricola, and Ercker — which continued to be made for over a century and a half after their first appearance. Though the scientific works of Descartes, Newton, and especially Hooke and Boyle are full of insights on the nature of metals, almost the only original writing on steel from a metallurgical viewpoint is contained in the following excerpts from La Fidelle Ouverture de l'Art de Serrurier *by Marthurin Jousse, published at La Flèche in 1627 and never reprinted. This was written as an outcome of the author's connection with the construction of the Jesuit college at La Flèche in western France. It was one of three books from his pen that between them cover virtually the whole craft of building in wood, metal, and in stone.*

Jousse's interest was not primarily in the materials themselves but in what could be made of them. Most of l'Art de Serrurier *is devoted to the design of locks, lock furniture, and the tools and techniques for shaping and finishing them. There are chapters on the equipment and organization of the forge and on the training of apprentices. Among the machines that Jousse describes are a crank-driven wheelchair and other devices for amputees, a file-cutting machine, a mill for rolling lead cames for glass windows, and a heavy screw for a press. Incidental metallurgical information includes the composition and use of hard and soft solders, the casting of soft metals in plaster molds, and the blueing of steel and etching designs thereon. As a good craftsman Jousse knew the importance of his materials, and he appends several chapters on the selection of iron and steel for various kinds of work and on the hardening and tempering of steel. The section here translated constitutes Chapters 64 to 68, on pages 137–48 of the 1627 edition.*

Jousse is of particular importance in the history of metallurgy because of his excellent description of the hardening and tempering of steel and for his discussion of tests for selecting iron and steel based on a careful study of surface characteristics and fracture. His treatise reflects well the differences among irons from different sources and even the variability of iron from a single forge, the inevitable result of the lack of understanding of the effects of heat treatment and variations in carbon content and minor impurities. He clearly understood that a given ore could produce either iron or steel depending upon how it was handled in the fire, in opposition to the popular impression that some ores gave iron and others only steel. The production of steel by carburization in the forge fire is described, as well as that by longer heating at lower temperatures in appropriate compounds, i.e., both the superficial carburization or casehardening of finished iron objects and the conversion of bars of iron throughout their whole section into blister steel for further fabrication.

Jousse records for the first time that a smell of sulphur is emitted by hot-short iron when it is being forged. He compares the behavior of mineral coal and charcoal as fuels. Every page bears evidence of having been written by a practical man describing what he knows from his own experience, with no theoretical picture to warp his observation.

The Selection of Iron and Steel, and the Manner of Hardening It by Quenching

How to Recognize Iron That is Soft and Can be Worked Cold

Since art is nothing without its material, and since the artisan, however skilled he may be at his craft, can produce nothing valuable or commendable unless he has a material or substance that is suitable for his undertaking, I consider it of the utmost importance to describe here the method by which good iron can be distinguished from bad, so that the smith, certain of his material, can practice and apply his art to it surely and confidently.

In order to select soft iron, it is first of all necessary to know from which forge it comes and whether the ore used was soft or brittle, although it may happen that iron from the same forge and from similar ore sometimes turns out soft and sometimes brittle, even from the same pig. Pigs are large pieces of [*cast*] iron, ten or twelve feet long, weighing fifteen or eighteen hundred pounds or more, and of a triangular shape, which, after originally being cast in sand [*at the blast furnace*], are then brought to the forge.[a] One

---

[a] *The operation referred to is fining, in which wrought iron was produced from a previously smelted cast iron. In chemical principle it was similar to the later puddling process, from which it differed in the form of the furnace used and the disposition of the fuel and oxidant. It was quite distinct from the earlier medieval bloomery, wherein the ore was reduced directly to metallic iron in spongy aggregation without ever having been heavily carburized or melted. In the seventeenth century the indirect process was growing in importance because of its efficiency, but there was still a good*

end is there placed in the large hearth, where the refiner heats it until it begins to melt, although he does not cast it again as in the beginning. When the end (as a guess, a piece weighing fifty or sixty pounds or more) is very hot, the refiner breaks it off and lets it fall onto the bottom of the hearth, where he lets it heat, turning it in the fire and throwing dry sand on it to prevent it from burning. When it is very hot, it is removed from the fire and subsequently placed under the big hammer which strikes upon it, lightly at first, to double it, weld it, and draw it out into a bar two or three feet

---

deal of iron made by direct reduction from the ore in the bloomery. In the finery hearth a piece of cast iron from the blast furnace was melted and then subjected to an oxidizing blast to remove the carbon, silicon, and some other impurities. Most of the action would be indirect, through the intermediacy of iron oxide-rich slag. The metal was maintained at such a temperature that, as the carbon was removed, the purer iron crystallized out of the liquid to form solid grains, which were gradually gathered together into a lump and then forged to remove as much as possible of the entrained slag. Good descriptions of the process can be found in John Percy, Metallurgy, Iron and Steel (London, 1864) and H. R. Schubert, History of the British Iron and Steel Industry, 450 B.C. to A.D. 1775 (London, 1957). Since the impurities were progressively concentrated in the liquid which decreased in amount at the end of the fining, it was easier to remove them by oxidation, and the granular crystalline material was of higher purity than if the whole mass had been liquid. In essence a simultaneous physical and chemical purification takes place — zone-melting plus oxidation. In the case of the uniform liquid steel of the Bessemer and Siemens processes, purification must carry the whole liquid to the low concentration of impurities which is desired in the final metal.

The modern use of the word steel covers almost any alloy of iron. In Jousse's day steel was simply an iron-like material that was capable of being hardened by quenching. Many people thought that steel came from ores that were distinct from those of iron, although Jousse correctly stated that an ore will give either material depending on the management of its smelting.

Apart from the obvious danger of enclosing blobs and stringers of slag (some of which was inevitable, and even desirable if sufficiently finely dispersed, when it gave a fibrous texture to the iron), the principal defects in wrought iron were of two kinds — hot shortness, due mainly to the presence of sulphur (which gives rise to a low melting point liquid that is carried by capillarity between the crystals of iron), and cold shortness, which was principally due to the presence of phosphorus and caused the iron to break with a brittle transcrystalline fracture. Though these phenomena were clearly distinguished at an early date, and related to characteristic differences in the details of texture of the fracture, their chemical explanation was not forthcoming until the end of the eighteenth century.

long. It is then allowed to cool. While the bar is being drawn out, the forge fire continues to burn, and the aforesaid pig gradually moves by itself farther and farther into the fire because the other end lies higher than the hearth. At times, when the iron that comes out of the fining hearth is too hot and boiling, the refiners throw on it small, almost powderlike, pieces of still unrefined iron. I believe that it is this [cast] iron or the sand that is thrown on which is responsible for those grains in the [wrought] iron which are so hard that they sometimes have to be removed with a chisel or burin.[b] When you know from what kind of ore the iron has been produced, you can judge it by the following means without breaking it.

Choose iron bars on which you see small black veins running lengthwise; these bars should feel flexible when they are handled and have no tears, if you can find any. Above all there should be no cracks on the bars, that is, no small transverse fissures. If these occur, it is a clear sign that the iron is hot short, which means brittle when hot, and that there will be trouble in forging. In order to ascertain better whether the iron is soft and ductile when cold, it must be forcefully thrown flat [138] on the pavement of the street. If it does not break then, it will be ductile. In order to make doubly sure, take a cold chisel made of good steel and with it notch the bar in a transverse direction at the place where you wish to break it. Then put it in a "breaker" (or over a hollow specially made in a piece of wood or stone, or on a wooden block on which two pieces of iron bar are laid six inches apart from each other) in such a way that the part of the bar that is to be broken is over the empty space. Then strike it with the peen of a heavy hammer. If you have no breaker, break it on the anvil.

[b] *The addition of dry sand was for the purpose of fluxing iron oxide and oxidized impurities, while the addition of grains of cast iron (which was not a common process) was supposedly for the purpose of correcting over-oxidation, or perhaps actually to harden the finished iron by introducing a small amount of carbon into it. This would be useful if it were given time to diffuse uniformly, but, as Jousse remarks, very undesirable if local hard spots of cast iron remained.*

How to Distinguish Good Iron from Bad after It Has Been Broken

Soft iron can be recognized by the color of its fracture. When this is black over the entire cross-section of the bar, it is a sure sign that you have good iron which can be formed cold and worked with the file; for the blacker the fracture, the softer will be the iron to the file and the more ductile. But it is likely to have cindery spots, by which I mean that it will not be bright and shining when polished but will have spots on its surface which look as if gray cinders were mixed in with it. This makes it difficult to polish it and give it a high luster. I do not wish to imply that this is the case with all bars that are so black, but it happens very frequently.

Other iron bars have a gray fracture, that is, black mixed with white. Iron of this color is much harder and stiffer on bending than the previous kind. It is very good for use by blacksmiths for shoeing horses and to make the tools used by edge-tool makers, and is used also by wholesale producers of blacksmiths' ware. But as far as working with the file is concerned, the iron is likely to have grains and spots which cannot be removed with the file. If one of these occurs in the stem of a key which must be bored or drilled, it prevents the drill from going in straight and will cause the key to break.

Still another kind of iron has a mixed fracture — part of it being white and the rest black or gray — and a somewhat coarser grain than the iron previously described. This is often the better iron. It forges more easily, is not likely to have cindery spots, has no [*imbedded hard*] grains, and polishes more easily. I believe that it is the best kind for forging, working with the file, or easy polishing; for it is refined by forging and acquires an entirely black fracture when it is being worked.

Still other bars have a very small grain, like steel, and are ductile when cold. This kind of iron is difficult to work with the file and boils in the hearth. Even though it is difficult to forge or file, it is very well suited for use by blacksmiths who produce tools for working the earth. [139]

How to Recognize Iron That Is Brittle When Cold

Another kind of iron has a coarse, bright grain in its fracture like bismuth or talc. This iron is practically worthless, for it is cold-short and weak in the fire, being unable to withstand great heat without burning. Furthermore you will find when you handle the bars that they feel stiff to the hand, and a bar that is thrown down on the pavement, as I have described, will break at three or four places at a time. Thus, this iron can neither be dressed nor worked cold, and some of it becomes more brittle when forged into small pieces than it was before it was reforged, which clearly indicates that the ore used for it was brittle or that the iron was melted and refined with freshly made charcoal just coming from the kiln. [140]

How to Recognize Hot-Short Iron Or Iron That is Brittle When Hot

This is recognized by cracks or fissures which run across the faces of the bars. Such iron is usually ductile and malleable when cold. Another sign indicating hot shortness is that, when being forged, such iron smells of sulphur and that, when it is struck, small sparks resembling little flames or fiery stars come out of it. When it reaches the color that is bad for it, which usually means a little whiter than cherry red, it breaks when hot, sometimes right across the piece. If it is hammered or bent while it is at this malignant color, it becomes full of tears. This describes the iron called hot short.

Iron from Spain is very likely to be of this quality and to contain grains which the file cuts only with difficulty.

All old iron that I have used that had long been exposed to the air or to evening dew was hot short.

There is no doubt that this is connected with some corrosive and biting quality of dew as experience demonstrates. It is a known fact that, when some part of the body is immersed in dew, it itches and sometimes even becomes chapped, which can only be caused by some biting quality which abrades the skin. It is not astonishing, therefore, that iron exposed to dew is changed and transformed and becomes, as I said, hot short.

We have in France some very good iron ores — if only they were carefully selected and cleaned and if, after being mined, they were kept for some time where they are exposed to the air; if only they were smelted and refined with charcoal made of young wood, prepared one or two years in advance, and kept in a dry place before the iron is melted and refined (because charcoal made fresh and of old wood makes this iron brittle, and the charcoal lasts hardly at all in the fire).

There are other ores in France which are used only to make iron; but they could furnish good steel if they were correctly managed and worked. I believe that this is much more a question of method than of quality, and that our mine and forge owners do not try hard enough to find and employ men who are capable and experienced enough to smelt and refine these ores and to know what kind of charcoal to use; for the role played by the charcoal is an important one, as I shall explain elsewhere. [141]

There are still other ores which, if they were skillfully sought, managed, and refined by people with experience, would prove of great value — if negligence would not dull the desire to find marvelous things to be relished with great satisfaction, if people with inquisitive minds would follow their instinct and investigate what mother nature secretly and gradually creates in her womb. This surely deserves painstaking effort; for things of value cannot be bought except with work.

How to Distinguish Good Steel from Bad

If there is anything where not only the locksmith but any other person who wishes to have anything to do with the forge and iron must proceed with particular diligence and care, it is in the selection and choice of steel. You accomplish nothing by having good iron and by knowing how to work and forge it if, at the same time, you do not know thoroughly and perfectly how to face it with steel.° Since nothing can be produced without tools and these are useless unless they are made of good, well-chosen steel, it is obvious that without steel and without knowledge of steel it is impossible to produce anything that is reliable and profitable.

In order to make a good selection from among the small pieces of our common steel (which is known as Soret, Clamecy, or Limousin steel, is the cheapest that can be bought in France, and is sold in the form of small squares approximately three inches on edge), it is necessary first of all to ascertain whether the squares have tears or overheated spots and whether black or straw-colored veins are visible in the fracture. All these are indications that a steel is not good. But if the squares are clean, without tears or overheated spots, and if in the fracture produced at the end of such a piece the steel is sound and the grain is white and fine, it is a sign that the steel is good.

There are other square pieces, which are bigger and weigh half again as much, but which are produced from the same ore as the little squares known as Clamecy steel. These must be selected in the way I have described. This steel and the little squares of Soret steel are good for working the earth and for heavy pieces of blacksmith ware.

Piedmont Steel

Piedmont steel is sold as square pieces which are a little bigger than those of Clamecy steel. A square weighs........ᵈ and sells for

---

° *Many tools were made of iron with an edge or face of steel, applied by welding to those places where the principal wear was to come. The process was known as steeling. Although it was done principally to avoid expense, it would generally give a more serviceable tool, combining toughness and hardness more effectively than in a tool made entirely of steel.*

ᵈ *Lacuna in original text.*

three *sols*, six *deniers*. In making a selection one must ascertain that the pieces are clean, that they have no tears nor overheated spots, which are recognized when there are places which are lumpy, have traverse fissures, and feel stiff when handled, all of which proves that the steel will be difficult to use and to hammer weld. Inspect the fracture to make sure that there are no yellow-looking spots, because this color is additional proof that the steel is difficult to weld and to unite with iron or another steel. [142]

But when the piece is bright and clean and when the grain is distinct, small, and white, and there are no black veins, and when it breaks easily at the quenched end when you strike it against a piece of iron or another square of steel, it is a sure indication that you have good steel, steel suitable for the production of the implements used for cutting bread, meat, horn, wood, paper, and the like, after it has been piled and welded as I shall explain hereafter.

Another Kind of Piedmont Steel

...There are two kinds of Piedmont steel, made by different processes. One is artificial and the other is a natural steel produced from a good ore, but it is the latter which usually has tears and overheated spots and a coarse grain of a pale color, and which is very difficult to weld. The artificial steel is the more common variety.[e] It is made of small pieces of iron, which are placed in alternate layers with specially prepared crushed wood charcoal in a large crucible or specially made fire-resistant pot, which has a lid and is covered in such a way that no fumes can escape. Subse-

[e] *This is the earliest good description of bars of "blister" steel made by the cementation process as distinct from short heating directly in the charcoal fuel of a hearth, though Porta (see preceding chapter) had described the box-hardening of armor and files. It is interesting that Jousse refers to cementation steel as the more common Piedmontese variety. The process is generally supposed to have developed in the low countries and Germany, and to have spread to England very early in the seventeenth century. (See Schubert, loc. cit., pages 321-30, who does not, however, mention Porta or Jousse.)*

quently this pot is placed in a kiln in which lime is burned or tiles, brick, or earthenware are fired or, better still, in a specially made furnace which is used for no other purpose.

This steel is of good quality provided that it has been refined twice and that the charcoal with which it was refined was freshly made shortly before it was employed. Note that not just any charcoal is suitable; do not go wrong here. The crucible must be exposed to a violent fire for two days and two nights, the longer the better, provided that it admits no air. This steel is good for working the earth and to steel-face hammers and other tools used for forcible and violent work. Under certain circumstances, it is also used for cutting tools, when it is thoroughly refined and properly quenched.

Steel from Germany

This steel is sold as small square bars, seven or eight feet long. It is very well suited for making springs for locks, the backs of crossbows, swords, springs for arquebuses, and other springs. In order to be good steel it must be clean, without tears or overheated spots, and without black veins or streaks of iron. All this can be detected in the fracture.

Carme Steel, or Steel with a Rose

We import this steel[f] into France from the states of Germany and from Hungary. It is very good for making cold chisels, burins, gravers, scythes used to cut grass, and for other tools for stones, horn, paper, and wood. This steel and the one described in the preceding section are among the best we use in France. It also is recognized as good if it feels flexible to the hand all along the bar and has no tears nor overheated spots; and if in the fracture,

---

[f] *For a discussion of* acier à la rose *see the note to the excerpts from Réaumur, pp. 71–72.*

right in the center, there is an almost black area verging on violet, with very fine grain and without any yellow spots or signs of iron, and if this area is found in the cross-section of almost the entire bar from one end to the other, these are sure indications that the steel is good. On the contrary, if the bars have tears and overheated spots, and if there are any veins in the fracture, the steel is not good.

Steel from Spain

We import from Spain big square bars, five, six, or seven feet long and eighteen or twenty lines square. This kind of steel must be selected like the ones previously described. If it is well selected, this steel is suitable for steel-facing anvils, bickerns, big hammers, and other heavy tools. [143]

Another Steel from Spain Called "Grain" Steel

We use still other kinds of steel from Spain, which are called *acier de Grain*, or else *acier de Motte* or *Mondragon*.[g] This steel is sold in big masses, in the shape of large, flat cakes, sometimes eighteen or more inches in diameter and two, three, four, or five inches thick. It is good for use as cold chisels, for steel-facing iron in crushers, hammers, and other heavy tools that must be hard and have to withstand much wear, and for cutting hard objects such as stone, marble, and the like, provided it is well chosen and well refined.

In order to indicate good quality, the fracture of the steel must have a fine grain and it must be almost completely yellow without black veins or any sign of iron. The workpiece must be taken from the center of the cake with as little as possible of the crust. If the grain is course and bright, with black veins and without showing

---

[g] *Mondragon, in Guipuzcoa, was a preferred source of steel for the exterior parts of the famed Toledo blades.*

any yellow, or if a piece from the rim of the cakes is taken, this steel will probably be worthless. Before it can be used and doubled and hammer welded, it must first be placed in a fire of charcoal or mineral coal. However, charcoal is to be preferred for this steel (as well as for the others which I have discussed heretofore) for mineral coal burns more violently and hotter than charcoal, and it is therefore not as easy to recognize when the iron or steel is hot because of the flames which surround it. This I have said before; I wished to repeat it here to be helpful to the reader.

After you have placed your steel in the fire and have heated it for some time, leave it there undisturbed for a short while and let it boil in the fire, sometimes throwing fine sand or smooth clay on it to cool it and prevent burning. After you have let it boil a short while in the fire, take it out and strike it as fast and as lightly as you can to flatten it and draw it into small flat bars, two lines or more thick. Then heat it until it is cherry red and plunge it in water. Subsequently, break it into small pieces and place them one upon another on an iron sheet, two or three lines thick, which has been covered with smooth clay tempered with water. Then heat the steel thus loaded on your iron sheet — and heat it gently. Afterwards you must nimbly remove it from the fire, striking it quickly and lightly, as you did before. When it has been thoroughly welded, draw it out to any thickness desired. It is possible to pile and hammer weld and refine the little squares of Soret, Clamecy, Piedmont, and other steels, and even to use them together and hammer weld one to another. This is sometimes done by cutlers and other experienced master craftsmen who know well how to use steel.

As far as Spanish and German steels in the form of bars are concerned — such as Carme steel (the Rose steel), and also steel from Hungary and other steel which is offered for sale as bars — they are not as frequently piled as steels sold in the form of squares, because they are not as commonly used for edge tools as Piedmont steel and others sold as square pieces.

Furthermore, in order to be of good quality and to be well chosen, all the steels of which I have spoken so far must be expertly handled in the fire; they must be protected against burning and overheating, which can be done in the way indicated.

But to forge his iron and steel expertly is not the most important task of the smith. He must know well how each kind of steel has to be quenched; and considering the product which he intends to make he must know how to find the steel suitable for his work. For not every steel is suited for making every kind of product. [144]

Dealing with Different Methods of Quenching Steel

We now come to that which constitutes the crowning accomplishment of the art — I mean the different methods of quenching iron and steel.[h] I hope that this will be as pleasing and agreeable to all smiths as it will be useful and profitable to each individual one. It seems that here we are dealing with one of the principal parts of the art because, even though it may be extremely necessary to select one's material judiciously, to forge it well, and to work it well with the file, nevertheless all this will amount to nothing or very little if a mistake is made in quenching. It is, therefore, the duty of a skillful and experienced lockmaker and smith to treat this part of the art with special regard, to choose the right kind of water, and to apply to this subject all the necessary artifice. I hope that those who understand and make use of what I shall say in this little treatise will derive a special satisfaction therefrom.

[h] *The word translated "quenching" is* tremper; *it could equally well have been translated "hardening" or "tempering," but we have preferred to avoid the uncertainty arising in the change of meaning that has occurred in the English word "temper" in the last few centuries. In the next section Jousse uses* tremper *to refer to an entire process of carburization followed by quenching; there we render it "hardening," although Jousse, unaware of the chemistry, thought of the operation merely as another quenching operation preceded by a special preparation in the boxes. Réaumur in 1722 still used the word in its old meaning, though he commented on the confusion.*

How to Quench the Little Squares of Limousin, Clamecy, and Artificial Steel

After your pieces have been forged, faced with steel, and dressed, heat them a little beyond cherry red; then quench them in well or spring water — the colder, the better. Some people put glass in the forge fire before the steel is heated therein and make it melt and coat their entire workpiece; they then quench it very hot. I believe that this accomplishes nothing.

Others take common salt, crush it, and put some of it on the steel when it is hot and ready to be quenched. I believe that this makes the steel harder and less likely to break. For this reason I use this practice with my own hammers, and with other similar pieces, to make them harder and give them more resistance to withstand the blows and stresses to which they are subjected.

After you have heated your steel and put salt on it, you will immediately put it in fresh water, as described, and keep it there until it is cold. If it seems advisable, you will subsequently temper it.

How to Quench Piedmont Steel

If you have made implements for cutting bread, meat, wood, horn, paper, or the like, they must be quenched cherry red and later tempered at such a heat that, when you pass a piece of dry wood [145] such as the handle of a hammer or the like over the cutting edge, the shavings or scrapings that come off the wood will at once burn on the piece. If this happens, it will be satisfactorily tempered. Remember that every steel that has been quenched too hot will be ruined and can never be given additional hardness. Many people do not agree with this. If steel has once been quenched too hot, it will always be worthless if you have failed to do it right the first time you quench it.

If it has not been quenched too hot but the tool does not turn out satisfactorily, you can quench it again and produce a better tool than by the first quench, as long as you recognize your mistake and know how to temper the steel. It is similar with all other kinds of steel, which you must be able to distinguish from each other before you can be sure of how to quench and temper each one.

How to Quench Spring Steel from Germany

The best and most natural of all waters is the dew of the month of May collected and squeezed out in the morning, at sunrise, at some elevated location, from cereal or other grasses, since it is less earthy, more subtle, and much more effective when it has been taken and pressed out of them at a time when all plants, roots, and herbs are at the height of their vigor. Its effect will be especially pronounced if it is collected or squeezed out when the wind blows from the north or northeast. For the coldness of the wind makes the dew more penetrating and effective, so that the steel that is quenched therein is left harder.

Take of this water six, seven, eight, or nine times the weight of your steel and put it in a vessel. Then heat the steel gently until it has become cherry red and take care that it heats evenly throughout, that no scale forms, and that it does not heat too rapidly. Then place it in the afore-mentioned water deep enough so that it cannot be touched by either draft or still air and let it cool there. Afterwards take it out and clean it with sand or ashes until it is white and all scale has been removed.

When your spring has been quenched and cleaned, put it back on the fire and let it slowly temper. It will immediately become yellow, blood-colored, violet, water-colored, and grayish-black. When it has reached this color, it must be removed from the fire, and a piece of dry wood must be passed over it, as I described in discussing Piedmont steel. When the scrapings or shavings of wood burn on it, take the horn of a sheep, a male or female goat, an ox, or any other oily horn and pass it over and rub it on the afore-mentioned spring; or pass over it a feather [*dipped in*] oil, candle grease, or any other grease. Then put the spring back again on the fire for a short while. When you have rubbed it with oil or grease, this must be allowed to flame up and burn on it, and then you will have to see once more whether the wood will burn. Afterwards the spring is allowed to cool, and you are finished.

It is quite feasible to quench springs in forge or river water as well as in spring or well water. But if you wish to quench them in spring or well water that is found to be too cold, you must pour it into a vessel in which you can stir and agitate it with a piece of wood or by hand. In this way you will soften the effect of the water even if it is ever so harsh.

If springs, or similar pieces, are quenched in well or spring water that has not been agitated, the spring will be likely to break either on quenching, if the steel is brittle, or else on bending. [146]

How to Quench Carme Steel, or Steel with a Rose

Heat your steel cherry red, using nothing but wood charcoal, and quench it in well or spring water; the colder and more penetrating the water, the better. If you are quenching a chisel or any other thin object, the steel is likely to break and split in the water. To circumvent this danger, immerse the thick end first in the water, the end that is the least hot and which is not intended for service. Lay it right down on the bottom of the vessel containing the water, or, as an alternative, pour molten grease on the water, which may be tallow or any other grease; then, after the piece which is to be quenched is hot, it is passed through this grease which floats on the water, and this will keep your tool from breaking. When it has been quenched, it must be tempered, after being cleaned as I have described, so that the temper color which you wish to give it can be more easily seen.

Tools for cutting iron, such as burins, gravers, chisels, and the like, are tempered at a yellow that has a little red in it and are then allowed to cool. If your tools splinter or break in use, place them back on the fire for a short while or upon any heavy piece of hot iron in order to temper them more. Let them, for instance, tend toward violet in color until they are of the desired quality. In this way you can make them as hard or soft as you wish, provided the steel is good.

Carme steel and steel from Hungary are also very well suited for making scythes for cutting grass or straw, and for making other tools. After such scythes have been made and dressed as necessary, they are quenched in a small trough or any other vessel which is as long as the scythe and deep enough for such tools to be completely immersed. It is customary to fill this trough with beef suet or any other grease, to which is sometimes added a small amount of [*mercury*] sublimate, arsenic, dragon's blood, copperas, verdigris, antimony, or rock alum. But I believe that these additions to the grease accomplish nothing. The scythes are quenched cherry red; then, depending on the quality of the steel, they are tempered violet or gray.

Scythes are sometimes quenched in the dew which I have spoken of before, mixed with rue and several other strong drugs and herbs, which accomplish nothing. Such water is capable of producing good tools provided the steel is good and it is properly tempered, which means tempered as described for spring steel. No other treatment should be tried.

How to Quench Steel from Spain

The steel from Spain that is sold in the form of heavy bars must be quenched like the Soret, Clamecy, and Limousin steels. Heavy pieces like anvils, bickerns, hammers, and the like are not tempered. They are quenched in their entirety in well or spring water — the colder and more penetrating the water, the better.

The other steel from Spain, which is sold as blooms, must be quenched and tempered like Carme steel, the steel with the rose. It has the same characteristics. [147]

How to Harden[1] Files and Other Tools Made of Either Iron or Steel

The best and most reliable method of hardening files and other pieces made of iron is the ordinary one in which chimney soot is used. It is important, however, to take the coarse soot which collects on chimney walls and is the hardest and driest kind that can be found, and not to forget to mix clay with the soot, which must be thoroughly crushed and pulverized so that it can be passed through a sieve. It also must be tempered with urine and vinegar, and a little common salt or pickling brine (which is dissolved salt) must be added. When all this is blended together, one has to be careful not to add too much urine and vinegar but to add it little by little and always to mix and grind hard. In this way only very little will be used to thin your soot, for the more you mix and grind, the more liquid the soot will become, and hardly any vinegar or urine will be necessary to thin it until it becomes as liquid as mustard.

After your soot has thus been blended, take a mixture of vinegar and salt and with it rub and scour your files by hand or with a cork in order to remove the grease which is put on them when they are cut. After they have been thoroughly degreased and scoured with the aforesaid vinegar and salt mixture, rub your pasty soot into them; make it enter into all the cut grooves of the file and cover them with it. Afterwards place them in an iron box or inside a hollow tile or something similar. If you have no box, you can make one of beaten smooth clay of the kind used in brazing. The aforesaid tiles are placed in the box in alternate layers with the soot. In the middle of the box is placed a tube made of iron or paper, which is as long as the files. It contains a test piece, that is, a little

---

[1] *The word here is* tremper (*quenching*) *as in the preceding chapter, not* endurcir (*hardening*). *Since, according to modern English usage, the two terms are not synonymous when they refer to the heat treatment described in this chapter, "hardening" has been used here. See fn. h, p. 55. In most other chapters in the present book* tremper *is also translated thus.*

iron rod placed inside the tube, which is to be withdrawn when according to your estimate the files will be almost hot. Place all your files in the box in this way, and cover them with soot. When they are all in the box, together with the soot, tie them up firmly in a piece of cloth, which must have been placed inside the box before the files were put in. Then you can easily tie up all the files with the soot by binding the corners of the cloth together with a string. After they have been thoroughly bound and tied up, the whole bundle will be covered with clay which has been beaten as in brazing, so that the files cannot be touched by any draft of air. Subsequently they will be placed in a wind furnace made of tufa, brick, or the like and heated with wood charcoal. They will be left there until they have reached a color somewhat beyond cherry red as if [148] you wanted to quench steel and produce tools for working the earth. You will recognize whether this color has been reached by means of the iron rod or test piece which is gently pulled out of the tube.

Small files made of iron must be heated and quenched much hotter than if they were old files, recut the second or third time, or if they were made of steel.

When you see that they are hot enough, they must be thrown into a vessel full of spring or well water, which will be the more effective the colder it is. If the files become bent in quenching, they can be straightened by bending them gently in the water before they are entirely cold and before they are taken out. If you wait to straighten them until after they are dry, you will break them in bending.

After they have cooled, they must be cleaned and scoured with wood charcoal or cork to remove the scum and soot which have remained in the grooves. When they have thus been cleaned, they are placed before the fire to dry until they are quite hot and all the moisture has evaporated. Then you put them in alternate layers with wheaten bran in a box or case to protect them against rust.

When you have smooth files, they must be wrapped or packed

in oiled paper lest the very fine flour contained in bran enter into the grooves of the files.

Another Way of Hardening Small Files, Taps, or Drawplates

If you wish to harden small files, taps, drawplates, and similar tools which do not have to be as hard and as rigid as those previously discussed, take old shoes or boots and wash and clean them to remove all dirt. Then allow them to char in the fire, and crush them promptly lest they be turned quickly into ashes. After they have been reduced to a powder, pass it through a sieve and mix it with vinegar or urine, or both, and add a little of the soot of which I have spoken before. Then pack your files in a box in such a way that they cannot be touched by a draft, heat them, and throw them into cold water as before. If they should become bent or distorted in quenching, straighten them in the same way.

Note that, if they are well hammered when cold before they are cut and hardened, they can be easily straightened, especially files used for slitting.

There are still several other methods of hardening, which I did not wish to describe, because they are not as reliable and easy to use nor as cheap as these.

CHAPTER FIVE

On Methods of Recognizing Defects and Good Quality in Steel and on Several Ways of Comparing Different Grades of Steel

René Antoine Ferchault de Réaumur

Extract from *L'Art de Convertir le Fer Forgé en Acier* (Paris, 1722). Reprinted from the English translation by A. G. Sisco (Chicago, 1956), pp. 176–204.

Editor's Note

The book by R. A. F. de Réaumur, from which the following chapter is excerpted, is a towering landmark in sidereurgical literature. Its author was famous for research in marine biology, entomology, mathematics, thermometry, and the manufacture of porcelain, and as an editor (albeit a dilatory one) of the Descriptions des Arts et Métiers *of the Académie des Sciences. Not only was this the first book of significant length to deal exclusively with iron and steel, but it was written specifically to divulge the science behind the ancient crafts of iron and steelmaking for the purpose of improving operations on a commercial scale. The most immediately valuable part of his work was the development of methods of making malleable iron castings. This was the initial aim of Réaumur's metallurgical studies, but a succession of unsuccessful trials caused his interest to turn toward the simpler problem of making steel. This led to the development of ideas on the nature of iron and steel in general and suggested experiments that in turn gave the annealing process for the production of malleable castings. The book, entitled* L'Art de Convertir le Fer Forgé en Acier et l'Art d'Adoucir le Fer Fondu. . . , *appeared in 1722 with the endorsement of both the Academy and the King. It is a long book, written in an exasperatingly verbose style. Of the twelve memoirs on the conversion of iron to steel and the six memoirs on cast iron, we have chosen the one dealing with the testing of steel, for in this the scientific quality of Réaumur's mind is best shown. It is a model of experiment and theory in symbiotic relationship in the best modern style. Many decades ahead of his contemporaries, he had a full apprecia-*

tion of the value of tests, both in production control and in selecting material for service. Particularly to be noted is the methodical way in which he exploits the appearance of fractures as a test of quality, as a characteristic with which to identify the various grades of iron, and above all as an aid to the development of a theory to account for the chameleon-like character of iron under various conditions. Réaumur's work is permeated with a sense of structure, for he was profoundly influenced by Cartesian concepts of the corpuscular nature of matter, which he had absorbed mainly from Jacques Rohault's Traité de Physique (Paris, 1671). These ideas had a great effect in the seventeenth century on advanced scientific thinking, but, since they were largely qualitative in nature and of unlimited flexibility, no quantitative use or proof of them was possible, and they gradually disappeared from the writings of the principal scientists. It was, however, taken for granted that matter was composed of some kind of parts, and the corpuscular concept eventually gave rise to the Daltonian atom and molecule as well as to the polyhedral boxlike cells that provided the structural units of the first mathematical crystallographers (see the paper by Grignon, Chapter Seven). Speculations on more complex aggregates did not return to metallurgical science until the end of the nineteenth century. Réaumur's work, therefore, quite failed to precipitate the revolution in metallurgical thinking that would appear to have been its logical consequence. Fracture tests continued to be important in practice. A richer sense of structure than that of Réaumur had to await the application of the microscope and the discovery that the grains visible in the fracture of metals were crystalline in nature. The advances in the intervening years were chemical rather than physical.

A short excerpt cannot do justice to the whole range of Réaumur's studies or his thought. One of his best sections is Memoir X on the examination and selection of steel, which is here reprinted in full. It is followed by two others on the hardening of steel, Memoir XI being more or less theoretical and XII, practical. Réaumur comments that *quenching is nothing more than a sudden cooling, and he rejects earlier theories in which something from the water was supposed to enter the steel. He confirmed Perrault's observation that quenching increased the specific volume of steel and proposed an ingenious structural theory to account for its increased hardness in terms of the movement from the body of the grains of steel into their interstices of the "sulphurs" that he believed to distinguish steel from iron. The following is a brief excerpt of his explanation:*

We know, and it is important to remember, that iron and steel soak up the sulfurs; but we also know that violent fire can remove them from the metals. For the present, in order to eliminate as many unnecessary factors as possible, let us look at a single grain of an unquenched steel, one of those visible to the unaided eye, and let us then compare it with a grain of quenched steel of approximately the same size. This grain (Pl. IV, Fig. 5, G), which is easily seen by the eye, is itself an accumulation of an infinite number of other grains, which we shall call the "molecules" of this grain ($M\ M$). The microscope brings these molecules into the field of our vision. But these molecules are themselves composed of other parts ($p\ p$). It is possible, if it seems desirable, to suppose that the latter are the elementary parts, although in reality we may have to continue the division vastly much farther before we reach them; however, we can stop here. We thus have to consider a grain, the molecules of which it is composed, and the elementary parts of the molecules. As the salts and sulfurs intimately penetrate the iron, we can at least assume that those by which steel outnumbers iron penetrate the molecules of the grain. If I expose to the fire a soft steel containing the grain on which we have concentrated our attention, the fire will melt the sulfurs and the salts of molecules of this grain before it melts the molecules themselves. It will drive part of the sulfurs and salts by which steel outnumbers iron out of the molecules in which they were wedged. Whereas, before this, they had penetrated those molecules, they will now, as a first step, go into the gaps between them. This will be all the effect caused by a moderate fire.[a] It is not just because it helps me to arrive at my explanation that I

[a] *Had Réaumur only reversed the effect of fire in this admirable theory, he would have precisely anticipated the modern theory: Slow cooling permits the carbon to aggregate (as iron carbide) in the spaces between the iron grains, while heating to a*

make this assumption. Reconversion of steel back into iron, accomplished either by a slow but long-continued fire or by a violent fire, has proved that the fire robs steel of its sulfurs and salts. Those which it has forced to come out of the molecules first occupy the spaces between these molecules. Let us therefore not hesitate to admit that, when our grain has reached a certain degree of heat, the empty spaces between the molecules of which it is composed will be partly filled by a sulfurous matter which was not there previously and of which the molecules have been deprived; that part of this sulfurous matter, which the fire has started on its way out of the iron, has passed from the molecules themselves to the intervals left between them. In this state let us plunge the bar of steel with the grain we are studying into cold water. We shall instantaneously fix the sulfurs and the salts which float around together. We shall deprive them of their fluidity; they will no longer be in condition to re-enter the molecules. However, the small intervals between these molecules of the grain will now be more completely filled, and filled by a substance which we can suppose to be almost as hard as we want it to be. The molecules of the grain will therefore be more firmly bound to each other. For this reason our grain of steel will be more difficult to divide or to break; in other words, our grain of steel has now become harder. The same thing has happened to all the other grains of steel that had acquired the same degree of heat. Consequently, our steel has now been hardened; or, to be more precise and in order to keep in mind what we actually wish to explain, we should say all the grains of our bar of steel have now become hardened.

As a good scientist Réaumur was concerned with measurement of the properties that he discussed. He tried to reduce the steelmaker's elusive "body" to a measurable degree of bending before fracture, in the instrument shown in Figure 1, Plate IV. In Figure 3 is shown a

---

temperature for quenching distributes the carbon uniformly in the interstices between the iron atoms. Réaumur's belief that properties of steel depended upon the relation between the main parts of the material and the salts and sulphurs (which today we know to be just carbon atoms) indicates a physical intuition of the highest order. It is curious that his approach was not made use of by his successors — indeed, the physical study of metals as opposed to the chemical or engineering study of their properties was long delayed.

hardness test, in which the mutual indentation of two crossed prisms produced by a standard blow is compared. Elsewhere he had proposed the use of a series of minerals for measuring scratch hardness, anticipating the better-known test of Mohs. Réaumur describes vividly the two main methods of hardening steel, namely the interrupted quench and the full quench followed by tempering, and the use of temper colors in their control. He discusses the influence of the quenching medium on the de-scaling of steel on quenching, necessary if the temper colors were to be observed on the steel during the interrupted quench. Finding that steel quenched in nitric acid was excessively hard led him to suspect some penetrating spirits from the acid (an observation that later led Lavoisier into a ridiculous theory of hardening). Finally Réaumur incorrectly attributes the sequence of temper colors to the "sulphurs" rising to the surface in various amounts.

Altogether, despite its prolixity, Réaumur's book gives a most valuable glimpse of both theory and practice during the transitional period in which he was writing and is essential reading in extenso for anyone concerned with the history of metallurgical science and practice. The whole book has been translated into English: Réaumur's Memoirs on Steel and Iron, translated by Anneliese G. Sisco (University of Chicago Press, 1956). It is from this that the following chapter is reprinted. Numbers in square brackets denote the pages in the original French edition. Our Plates II, III, and IV are respectively 8, 9, and 10 in the original.

On Methods of Recognizing Defects and Good Quality in Steel and on Several Ways of Comparing Different Grades of Steel

STEEL can be so bad that it is recognized as such at the moment of inspection; it can have defects which are obvious enough for that. I have repeated more than often enough that, when there are many cracks (Pl. II, Fig. 1) or when the corners of billets appear jagged here and there (Fig. 2), one can count on it that it will be very difficult to process the steel. In our country such defects can be considered to be a bad sign even if they are extremely fine. But is it entirely the fault of the steel when it is found to be so bad, or is it perhaps a little the fault of our artisans? This question should at least be asked, and the answer will perhaps not be favorable to our artisans, if one expects from them as much skill in treating steel as the artisans of the Levant possess. His Royal Highness the Duke d'Orléans, himself, gave orders that some of the steels of which the famous Damascus swords are made be sent to me from Cairo. M. le Maire, who was consul at Cairo at the time, responded to this request to the best of his ability. [260] Among the steels which I received from him, and which he assured me were the best, there is a cake which is supposed to be steel from India and the kind to be rated most highly in Egypt. I could find no artisan in Paris who succeeded in forging a tool out of it. It withstood the fire hardly better than cast iron. Other steels from the Levant are ordinarily difficult to forge but less so than this one.

If the surface of a billet is full of laps (Fig. 8, *M*), one should not expect to be able to forge it into sound products. When the steel is sold, it has been quenched and broken; if veins of iron or spots with non-uniform grain — that is, numerous brilliant platelets mixed with gray or dull grains — are noticed in the fracture, the steel contains much iron and consequently is no good.

But the defects of steel are not always so easily recognized. A bar or billet may appear very sound and nevertheless belong to the kind that is hard to work. The test that will really show if steel is without this defect [of being non-workable] is to forge it after it has been given a welding heat. This means that it must be given the heat necessary to force the two ends of a bar, after one has been doubled over the other one, to be united by hammer blows perfectly enough to give the impression that both of them are one and the same body. This is what the smith calls "welding." A bad grade of steel can still look very sound if it has been forged at red heat; but it will be full of cracks if it has been heated [261] until it was whitish-red or if it was brought almost to welding heat. It would be impossible to weld it without giving it almost this latter degree of heat. With a bad steel, the two ends that are to be welded together will not weld perfectly; but when good steel is broken at the location where it has been welded, it will be difficult to recognize where the two parts once were separate.

Even at the time of heating steel to the degree necessary for welding, it is possible to predict whether it will be able to stand as much heat as it must be given. One only has to listen. A noise is heard which can be distinguished without much practice from the noise made by the flame or from that of the blast of the bellows. It is the result of something like boiling which takes place in the steel. Steel that boils this way will not forge well. Another bad sign concerning the quality of steel is observable while steel is in the forge; this is when its surface continues to look dry, when something like a gloss does not appear on it after it has been heated thoroughly and has even been sprinkled with some sand.

When well-heated steel is put on the anvil and breaks under the hammer, or if it splits or breaks at least partly in the bend when one end is bent over in order to weld it on to the rest, such steel will not be easily workable (Fig. 3, *D*).

The smith should beware, however, not to blame the steel for defects which may be caused by his own negligence [262] or ignorance. Good steel can be made bad by excessive heating, by causing it to melt or almost to melt.

One indication of a good steel, which the merchants of Paris mention most boastfully and to which much importance is attributed by artisans of mediocre skill, is the rose.[b] What they call

[b] *The rose seems to be a transverse fissure which has become somewhat oxidized to show temper colors. Its appearance must depend on many factors, and it is indeed astonishing that it should ever have become a mark of anything but inferior quality.*

A century earlier Jousse (La fidelle ouverture de l'art de serrurier [La Flèche, 1627]) had remarked that the best steels that were used in France were those imported from Germany and Hungary under the name of "aciers de Carmes ou à la rose." They were recognized, he said, by the fact that "the bars could be bent by hand, were without laminations or burned spots, and showed in the middle of the fracture a nearly black stain, tending toward violet, having a very fine grain, flawless and with no patches of iron. If this stain covers practically the whole of the bar reaching all sides, it is a sure sign that the steel is good. On the contrary, if the bars are laminated, overheated, and with some veins [of iron] intermixed in the fracture it is not good."

A good description and essentially accurate explanation of the rose was given in 1771 by Jean Jacques Perret. This was in his L'Art du coutelier *published under the auspices of the Académie des Sciences as one of the* Descriptions des arts et métiers. *We translate in full the section relating to the rose: "In the old days a German steel which is called* Carme *steel was very good; but it has now deteriorated, at least as steel for razors, although it is still very good for making knives. It often displays the rose. This is mistakenly considered to be a mark of quality. If the middle of the fracture of a bar of Carme steel is blue, black, or violet (which is what is meant when we say it displays the rose), it is an infallible sign that the texture of the steel is internally broken. Experience has proved that this rose is not found throughout the whole length of the bar and that a light hammer blow will cause the bar to break at the spot where it displays the rose, whereas a much heavier blow is required to break it at some place where the rose does not occur. I believe that this rose has its origin partly in the quench and partly in the process of making the steel, in which connection it should be stated that this is a natural steel, which means that a*

the "rose" is a certain spot found in the fracture of some bars of steel. This spot can be compared with a rose only when it is round, but it is often oval and frequently has other, very irregular outlines (Figs. 4–7). The center of the fracture often is the center of such a spot. There are some that occupy the greater part of the fracture and others that occupy only a rather small part. Their color is as variable as their shape. Some are a rather light blue and others are dark blue, almost black; there are yellow ones of different shades of yellow; some are just a little duller than the rest of the fracture. Finally, there are some where these different colors form concentric bands. In short, the color of the roses of steel may be like any of the colors that can be given to the surface of steel or clean iron heated gently over charcoal to be either colored or tempered. I shall discuss these colors elsewhere, and, when I explain what causes them, I shall also give the reason why the spots called "roses" [263] are produced inside of some steels. They vary as much within the same bar as among different bars. A bar broken

---

cement is not used in making it, but only 'boiling.' Furthermore, I will say, to explain the formation of the rose, that certain bars are quenched much hotter than the strongest cherry-red and are plunged suddenly into very cold water. The outside thus cools suddenly, while the inside of the bar is still hot. Even if the inside takes much longer to cool than the outside, and the contraction of the parts has to be proportionate to the cooling, both inside and outside must eventually shrink to the same extent. But, as the outside has suddenly felt the full effect, it has been left in the condition which it must reach; then, when the inside finds itself compelled to contract slowly, the grains are forced to break or spread apart in order to yield to the force of contraction. During this operation internal cracks are formed, or, at least, the parts are not brought close together at certain spots, because the outside, which has also shrunk, has operated first. This is, I think, how the rose is formed, and the best proof I can give is that, if the end of a piece of steel displaying the rose is inspected under a light [sic; lampe *misprint for* loupe, 'lens'?], *it will be noticed that all around the rim, to a depth of 1 or 2 lines, this steel is white and has a very dense texture, while the rest is blue or violet or black and of a much looser texture. I believe, therefore, that the occurrence of the rose is not as important as certain people would like us to believe. Good cutlers ought not to be guided by such an unreliable sign in choosing the material on which their reputation depends. They should not let themselves be seduced by the talk of merchants who practically always ignore that which constitutes a good steel and who, desiring to make a sale, attempt to make the bad qualities of their merchandise pass as perfections."*

at one location will have a rose of different color, different size, and different shape from a rose found when the bar is broken at some other location. The bar may even be broken where there is no trace of a rose.

There really is no more unreliable sign of good quality of steel than this rose. It is true that it is found in some steels that come from Germany and are very good; but it is also found in mediocre steels. Some of the steels from the Champagne and Limousin sometimes have it, and the greater part of the good steels from Germany do not have it. I have encountered it in steels from Foix (Fig. 4), which are so coarse that not even the people who produce them consider it right to call them "steels"; they call them "hard irons." What the rose indicates most reliably is that the steel in which it is found is steel produced from pig iron. I have never seen it, at least not well marked, in steels produced from wrought iron.

Furthermore, this spot is not permanent. It depends on the degree of heat the steel is given before it is quenched. When a bar with a rose is heated very hot, the rose will disappear in the first heating or will at least become fainter. It will not be found in the finished product.

In some of the provinces of our country where steel [264] is made from pig iron, it is forged into small billets, one end of which is hammered out into a point (Fig. 8). This pointed end (F) is quenched and is used in a test, of sorts, of the quality of the steel of the bar. To make this test, the bar is held in one hand, approximately at the middle, and a rather light blow is given the thick end (L). When the steel is good or at least reasonably so, the end of the bar opposite the one that has received the blow — the pointed end — will break. This test proves that there was no sizable vein of iron at the place that broke; but this is really all it proves. However, it means something if no vein of iron is found in the steels of this category.

If we except the afore-mentioned rough tests of which I have just spoken, the average artisan has almost nothing to guide him concerning the quality of the steels he wants to process. He does not know if his steels are suitable for the tools for which he selects them until the tools are finished, and in other cases only when those for whom they were made have employed them. Some honest cutlers, who belong to the more skillful in their profession, have repeatedly confessed to me that they did not know if a new steel which they had used for razors was suitable until the barber had given them a good report on the razors they had sold him.

The most valuable information concerning the steels with which they work is to know from which country [265] they come. The cutlers of Paris recognize as good only those from Germany. If they are offered steels from England or Italy, just because they are from England or Italy they will not buy them, even though they may be better than those from Germany. They work only according to routine. They are accustomed to certain shapes of steel. If the same steel is given a different shape, they will reject it, because they have no means at all by which to judge its qualities.

It would be especially important to have means by which to judge the merits of different steels and to decide for what purpose one kind is more suitable than another. I shall suggest some possibilities that have occurred to me; but I hardly expect that the common artisan will bother with them. However, they may at least be useful to those who try to distinguish themselves in their profession. Specialists, and scientists who wish to have certain products made that require carefully selected steels, can have the selection made according to these rules. They will be especially useful in steelworks; they will provide the makers with a means to be sure of the properties of the steel produced.

Thus, the difficulty here does not consist in detecting the defects that are more or less common in average steels. Steels in general have certain qualities which make them suitable for products that cannot be made of wrought iron. The difficult part is to determine the degree of perfection reached [266] by each of these qualities in every steel. The three principal ones are (1) that of acquiring a certain type of grain after having been quenched at a certain degree of heat, (2) that of being more or less hardened according to the degree of heat at which a steel has been quenched, and (3) the tendency to preserve more or less body after the quench.

The word "body" is ordinarily used to describe the ability of a quenched steel to resist the force that wants to break it. Of two steels of equal hardness which have the same resistance to the file, the one that is easier to break is the one that has less body. Of two flat bars of equal thickness which have been quenched in the same way, the one that is less rigid, that can be bent more, has more body. Of two chisels used to cut iron cold which have been quenched in the same way and which are equally hard and have edges of the same shape and size, the one with more body is the chisel that chips and breaks less easily while being hammered to cut the iron or some other material. It is in this sense that I am using the word "body" and shall use it in the future. It will always be descriptive of some kind of toughness which the steel keeps in spite of the hardness it acquires when it is quenched. The artisan who is little accustomed to connect clear concepts and well-defined meanings with his terms sometimes uses the word "body" to mean hardness. When the edge of a chisel folds back when it is used to cut iron cold, the artisans say that "it [267] has no body"; but I shall avoid using the word in this sense, which is improper. I shall say instead that such a steel "lacks hardness." A lack of hardness is the reason why its grains can be compressed and brought close together. We also hear it said that a steel lacks body when it does not readily withstand violent heating, when it becomes too weak therefrom. Of these meanings and still others which are sometimes given the expression "body of steel," we shall reject all but the first one.

If we assume that a steel has no laps, no cracks, no overheated areas, no veins or platelets of iron, and that it is easy to work, we must, then, examine three things in order to decide about its worth and about the applications for which it is best suited. These are its grain, its hardness, and its body. We must look for methods of determining if one steel is better than another in any one of these properties. For example, of two steels, the one that can be made to develop the finer grain will also be called the "finer steel." We still have to determine how these three properties are combined in the same steel, for we must realize that a steel loses body and fineness of grain as it gains hardness. The further problem is, therefore, to decide in which steels these combinations are the most favorable. However, before we go on to the combinations, let us review each of these principal qualities separately.

The fineness of steel or, differently expressed, [268] the fineness of grain that can be attained in a steel seems to be the most easily recognized of these three properties; it also can be most easily compared. Our eyes will make the decision. A good artisan admits, nevertheless, that by looking at the fracture of a quenched steel he does not know if it is finer than another. His reasons are very plausible. Heat a piece of very fine steel almost to white heat, quench it in this condition, and break it. Its fracture will show nothing but a coarse grain. Heat a mediocre steel, or even an agricultural steel, only to cherry-red and quench it. When it is broken after the quench, it will have a fine grain. The fineness of the grain thus seems to depend on the degree of heat a steel had when it was quenched. It actually does depend on it; but the important thing to know is whether, of two steels which have been quenched at the same degree of heat, one may not have a finer grain than the other. This is how excellent steels should be — and actually can be — distinguished from mediocre ones. However, it is not easy to decide by ordinary methods which of the two steels is the better one, because it is difficult (1) to heat two pieces of steel equally hot and (2) to break these two pieces in exactly the spot where they were equally hot when they were quenched.

In order to recognize better of what these difficulties and the inadequacy of the ordinary test consist, it should be [269] noted that a piece of steel that is being heated does not acquire the same degree of heat over its entire length; that, if it is heated at one end, this end may even start to melt (Fig. 8, F), while the other end (L) is still cold or at least black. When this steel is quenched and broken[c] near the end that was heated until it was almost melting (G), the grain will be coarse (Fig. 9, fg). When it is broken at a location a little farther away, the grain will be found to be less coarse (Fig. 9, gh), when it is broken still farther away, it may be found to have a fine grain (Fig. 9, hi); and when it is broken a fourth time, at a location still farther removed from the hot end, the fracture will have a much finer grain (ik). If another piece of steel which is to be compared with the first one is treated in the same manner, we may perhaps be able to decide which one of the fractures of these two steels represents the most beautiful grain; but we will be unable to decide which of the two steels really has the more beautiful grain. There is reason to wonder if the steel that appears to be inferior might not have outranked the other if it had been broken farther away from the end or if it had been broken at the middle of one of the pieces that were obtained by two successive fracturings. This doubt is especially justified, because the finest grain sometimes occupies only a very short length of the entire piece of steel.

[c] *This test was still in use two hundred years later. To quote from Harry Brearley,* The Heat Treatment of Tool Steel (*London, 1916*), *p. 90:*

The following means of determining in an ordinary smith's hearth the degree of redness, at which the finest-looking fracture is induced in steel by quenching, is generally attributed to Metcalf. A bar of the steel being used of about one-half inch diameter is notched for a length of three or four inches, each eight or ten millimeters from one end. The notched portion is then heated in the smith's fire until the extreme end is white hot, and in such a manner that the heat tapers gradually backwards. After quenching, the bar is dried and broken at each successive notch. The first piece or two will break off very easily — if they do not break in quenching — and exhibit a coarse, staring white fracture. Subsequent pieces will break off less readily, and become gradually finer in fracture until a smooth amorphous appearance of well-hardened steel with a faint bluish-grey tinge is reached. After that the fracture gets coarser down to the unhardened part.

William Metcalf described and illustrated his test in Metallurgical Reviews, *1877, 1, 245-53.*

When one end of a piece of steel is heated until it almost melts while its other end stays black, we can be sure that this steel has undergone all the different degrees of heating possible; for from the melting-hot end, which is the upper [270] limit of heat to which steel is susceptible, to the other end, the degrees of heat diminish very gradually, and the heat at the end that remained black may be considered to be zero, because it was unable to produce any effect. It follows that, when this steel is subsequently quenched, it undoubtedly is given all the different grain sizes it can have, as the degrees of fineness vary with the degrees of heat. In order to obtain a complete range of grain sizes, in order to observe all the degrees of fineness, and in order to make a comparison between two steels, it would therefore be necessary to break the steel lengthwise. This is impossible if the steel is left in its ordinary condition. Hammer blows break it transversely and at the location where it has been quenched most drastically, which is the location where it is the weakest.

To increase the accuracy of the comparison I have in mind, we have to concentrate, therefore, on eliminating the difficulties that prevent us from breaking a piece of steel lengthwise through the quenched part. There are several methods by which this can be done, all so simple that to find them only requires to decide to look for them. I shall suggest two from which to choose. The first is as follows: Of the steel which I wish to test, a flat bar is forged (Fig. 10). Its thickness and width can be chosen at will. A length of 2 inches is enough, but if it is 5 or 6 inches long, it will be even more convenient. Then I have this steel bar welded to an iron bar (Fig. 11) of at least the same [271] length and width but of greater thickness, if desired. When these two bars are thoroughly welded together, I take a cold chisel and cut through the iron bar down to the steel over its entire length. In the steel I cut a groove, something like a channel or furrow, which is then still widened a little by a file (Fig. 13, *g h h i i k*). That is all the preparation needed, the result of which is probably already anticipated. When one end of this

ON TESTING STEEL

bar, which is part steel and part iron, has been heated almost melting-hot and the bar has been quenched, it can be broken in two lengthwise; there is no reason to fear that it might break transversely. At the location where it has been quenched the hottest, and where the steel is most brittle, this steel cannot break transversely, because it is joined to the iron; it forms one body with it. The iron, which is not at all brittle, supports the steel. This is not the case at that part of the bar where the groove has been cut. All along this groove the steel is no longer supported by the iron, and the steel can therefore be broken longitudinally but no longer transversely. However, if the steel bar is several lines thick, and if such a long one has been chosen that a large part of it has not been heated and therefore is not in a quenched condition, the steel itself will be difficult to fracture in this [unaffected] area, as unquenched steel is hard to break. It can be done, nevertheless, by putting in the groove, at the locations that resist the most, a big, [272] somewhat blunt, chisel, something like a wedge (Fig. 14, *n*), which is hammered to force the steel to open up. Each of the pieces which has thus been separated from the other will show the whole range of different grain sizes (Fig. 15, *o*, *1*, *2*, *3*, *4*, *r*). Getting ready for the preceding experiment is in itself a test of the steel, as it shows whether the steel can be easily welded to iron.

There is still another method, which is easier than the previous one. One side of the piece of steel that is to be tested is left its original thickness; the thicker this side is, the better. The opposite side is hammered out over its entire length. To give a picture of the shape I wish the piece of steel to have, let me say it shall be forged out to resemble a blank for a razor (Pl. III, Fig. 1) with the exception that longitudinally its width shall be the same all over. The experiment will be the more successful, the thinner the hammered-out side is made or the thicker the opposite side has remained. The steel is heated melting-hot at one end and is subsequently quenched. Nothing remains but to break it, which is done by gentle blows upon the thin rim, from which different fragments are thus successively broken loose until the piece has been fractured over its entire length. It will make the fracturing easier if, before the steel is quenched, a groove is cut along the line where it is to be broken (Fig. 1, *B C*) — or at least all along that part which will not become hot enough to be brittle after quenching.

In the fractures [273] obtained by one of the methods just described, one glance reveals the whole range of different grain sizes a steel can be given by all the different degrees to which it can be heated. We are now able to compare the finest grain of one steel with the finest grain of another. Artisans who wish to know what they are doing should have a stock of such fractures of every grade of steel known in the Kingdom. These would serve them as standards to measure the degree of fineness of the steels which are offered them for sale. These bars would be for them what the touch needles (made of gold and silver mixed in different proportions) are for goldsmiths who must decide quickly about the fineness of gold that is offered them. Once made, these fractures will serve an artisan all his life, provided he is careful enough to put the bars in a dry place and to keep them in a box. I have had some for several years, the fractures of which are almost as white and brilliant as they were the moment they were obtained. I have taken the precaution, however, to put them in a box which, itself, is kept in another, much larger one, which is filled with bran.

Without a certain sensitivity of vision and, moreover, without some experience in inspecting steel, it would be hard to decide which of two steels of not too drastically different qualities has [274] the finer grain. But our fractures enable unschooled eyes to make a decision and provide a method of measuring the fineness of this metal by dividers. To explain this method, and to show upon what it is based, I shall demonstrate that the different grain sizes observed in the fractures (obtained by one of the methods just explained) can be divided into four types which are very easy to recognize. The first division contains only white, brilliant, coarse grains (Pl. II, Fig. 15, *o, 1*; Pl. III, *L, 1*); no other kind is seen, and we therefore refer to this type as white, brilliant [or coarse] grain. In the second division the white, brilliant grains are mixed with dull ones (Pl. II, Fig. 15, *1, 2*; Pl. III, Figs 2–6); but here the brilliant grains are less coarse than the first grains of the first division. We call this type the mixed grain. The third division has only one kind of grains, which are all fine, dull, and often gray (Pl. II, Fig. 15, *2, 3*; Pl. III, Figs 2–5). This is the fine grain. The fourth division begins where the preceding grain ends. Here we again find much coarser grains which, though dull, are less dull than those of the preceding division (Pl. II, Fig. 15, *3, 4*). These grains are ill defined; they do not seem to be quite detached from one another; they are, one might say, not lined up. This is the location where the steel has not been hot enough to respond to quenching. We therefore are interested only in the first three divisions. The grains in each of them are not of the same [275] size over the entire extent of the division; they become smaller as they approach the next order. Near the location where the mixed [or rather, the coarse] grain ends, where the second division just begins, the white, brilliant grains are less coarse and also less brilliant than the first grains of the first division. Similarly, among the mixed grain, there are more white grains and fewer gray ones toward the location where this order starts and, conversely, more gray grains than white ones where it ends. Thus I might have divided this order into two different ones if I had not been afraid to increase the number of divisions needlessly. The third division consists of much more uniform grains.

It follows from the foregoing that the dividing lines between the types are not very sharply defined — that they have to be approximated. Astronomers have to do the same when they must define the dividing lines between the umbra and penumbra of an eclipse and between the penumbra and full light. However, the dividing lines we have to deal with can always be defined accurately enough for our purpose. Furthermore, they can be defined with some precision when the eye is aided by a strong lens. The first grains of the fourth division are sometimes so fine that they would seem to belong to those of the third division (Pl. II, Fig. 15, 3, 4) — that means to those produced by quenching. But when they are viewed through a lens, when they are enlarged, one recognizes [276] the difference between them and the preceding group without the possibility of any doubt.

From these remarks, which I might have omitted if they were related only to the physical science of steels, and from my observations concerning the length of the divisions occupied by different types of grain in different steels, I am able to derive the promised rule for assistance in judging the fineness of every steel. I have observed that, as steels become finer, the extent of the third division becomes greater — it becomes longer compared with the second division. This difference is sufficiently considerable in steels of very different grades, so that we have no reason to fear that we might make a mistake in measuring them; we do not have to worry that we might misplace the border lines between the groups. I have seen bad steels from Nivernais in which the space occupied by the fine grain was not half as long as the mixed-grain division. On the other hand, I have seen fine steels in which the fine grain occupied more than twice as much space as the mixed grain (Pl. III, Figs. 6, 7). Consequently, these fine steels had at least four to five times as much fine grain as the coarse steels.[a]

If the fine grain of both grades of steel has the same degree of fineness and the same hardness, the one with the larger area of fine grain should be preferred for making tools which require this type of grain. It will be easier to obtain this grain in the cutting

---

[a] *The control of grain size is an important aspect of modern steel production, and steelmaking practice is adjusted specifically to increase the range of temperature over which proper grain size can be maintained. Réaumur's test would also show the effect of varying carbon content, for part at least of his second type of fracture would be due to ferrite plus austenite structures in hypoeutectoid steels at the time of quenching.*

edge of tools made from such [277] steels. For, if of two steels one shows in its fracture a larger area of the same type of grain which the other one has, it follows from the manner by which we heated these steels that this grain can be produced in the first by a greater number of degrees of heat than in the second. Thus, when we heat a steel (after it has been forged into a tool) in order to quench it so that after quenching it will have a certain definite type of grain, we can be much surer of succeeding in obtaining this grain if the steel can acquire it at different degrees of heat than when it acquires it at only one degree of heat. This degree of heat is difficult to attain exactly, and more so than one would think. Furthermore, we can rarely be certain that we have obtained it in all parts of the cutting edge. The parts which have become hotter will not have the grain size we wished to produce.

Finally, in the fractures of two steels, of which one is finer than the other but which have been heated equally over an equal length, the range of the quenched structure is greater in the finer steel. All this logically follows from our observations concerning the nature of steel. The finer ones, which acquire more [quenched] grain and finer grain, are composed of more detached parts which are richer in sulfurs and consequently of parts that are more easily set in motion because of these two factors. [278] There would be nothing better than this method of measuring the fineness of steels by dividers (one might call it) if doing it were as simple as one would like it to be. However, it requires great care on the part of the person who wishes to employ it in practice. It is necessary that the steels which are to be compared be heated to the same degree over the same length, and it is very difficult to accomplish this with ordinary wood or charcoal fire. Later in this memoir I shall give the means for overcoming this difficulty, which will once more become a problem in determining the body of steel.

If the steel that has been fractured is thick and was heated more on one side than on the other, the grain size at a particular location on the fracture may not be the same on both sides; but we can still observe the same sequence of divisions on each side; the relative position of each division will still be the same.

The coarse grain, the first division in the series, enables us to judge the fineness of steel at least as well as the fine grain does. Steels which have a finer grain than others in the third division often have a much coarser grain in the first. Something that is less subject to variation is that the coarse grains of this first division are whiter, more sparkling, and more brilliant in fine steels than in coarse steels; [279] they are also more uniform. The coarse grains of coarse steels seem to be mixed with dull grains. The first grains in the coarse division of a fine steel may at times also be dull if the end of the steel has been almost melting-hot; but the subsequent grains will have the most sparkling luster. The entire coarse-grain division of coarse steels resembles the beginning of this division in fine steels that have been heated until they were melting-hot; and sometimes the grains are even duller. Furthermore, in coarse steels the grains of the group under discussion sometimes seem to be arranged in layers. A similar arrangement is never found in the same type of grain in fine steels.

Let us now consider the methods by which the degrees of perfection of the second property of steel, its hardness, can be recognized and compared. All the difficulties encountered in judging the degrees of fineness are again encountered in judging the degrees of hardness. The same steel is more or less hard, depending on whether it was more or less hot when it was quenched. In general, the hardness decreases as the grain becomes finer. The location where the grain is the coarsest, the location where the steel was quenched the hottest, is ordinarily the hardest. Nevertheless, there are occasional exceptions to this rule, and these exceptions are enough to make trouble for those who must decide where a certain grade of steel becomes hardest on quenching. However, something which nobody has attempted to [280] determine, but which is nevertheless important, is which one of two steels that were quenched at the same degree of heat is harder. We have absolutely no rule on that. Again, our fractures furnish the means for formulating rules concerning this problem. These rules are as accurate as one could wish them to be and, what is more, independent of the difficulty of heating the steels uniformly before they are compared. If we wish to know at which degree of heat a steel must be quenched to acquire the greatest hardness, all we need know is with which order, or type, of grain it is the hardest; for the different types of grain are proportional to the different degrees of quenching. In order to compare the hardness of two different steels, all that is necessary, therefore, is to compare the hardness produced in each at its different divisions of grain size, and this comparison is easily made on our fractures.

The artisans know only two degrees of hardness — that at which the file takes hold and that at which it does not cut. These two expressions do not even come close to informing us of the degrees of hardness different steels can have. There is practically no steel — or, actually, there is no steel — which, after having been quenched at a certain heat, does not resist the file. We therefore must have files with a larger number of different degrees of hardness than ordinary files possess. Several kinds of stones or other hard materials serve me [281] as such files.[e] I use seven different kinds. The first is glass, which is less hard than ordinary files are. The second is the softest rock crystal. Pieces of hard transparent crystal, such as the crystal of Médoc, serve as the third kind; pieces of agate or a certain kind of faceted stone found near Perpignan as the fourth; oriental jasper as the fifth; oriental topaz or another of the transparent oriental stones, such as the sapphire, as the sixth. The diamond is the seventh and final one. I give these stones only as an example; everyone can choose others according to his liking and can increase or decrease their number depending on whether he intends to go more or less far with his tests. As long as the stones

---

[e] *The mineralogist will recognize here the anticipation by nearly a century of Friedrich Mohs's scale of hardness, which consisted of ten minerals, the hardest being (in decreasing order) diamond, corundum, topaz, quartz, and orthoclase.*

differ from each other in hardness, it does not matter. Glass may be followed by the ordinary file, and emery may also be given a place among the testing stones.

By means of these different files, by means of our different stones and our fractures which show all the different types of grain, we are able to determine and to compare the hardness of steels with accuracy. All that is necessary is to find out for each grain size with which stones one can make an impression, with which stones it can be scratched. The test is not made on the fracture itself but on one of the flat surfaces of the bar very close to the edge of the fracture. A few examples will make it [282] clear how this method works. If I successfully try to scratch with glass, or the file, the third type of grain, the finest grain of a steel, I conclude from this fact that this steel is not hard enough at the location where it has the fine grain — that no hard tool could be made of it with the degree of quenching that produced this grain, especially as I have seen that in other steels the same type of grain cannot be scratched even by emery. If I must make tools of very great hardness, to which the fineness of the grain is of less importance, and if I have found that the mixed grain of a particular steel cannot be scratched by agate or jasper, whereas the corresponding grain of various other steels can be scratched by rock crystal or the crystal of Médoc, I shall select the steel that resists agate. Finally, if I need extraordinarily hard tools, and I have found by my tests a steel (a kind which does exist) whose coarse grain resists oriental topaz, although other steels can be scratched by jasper or even crystal where they have the same type of grain, I shall choose the steel which can resist topaz. This is enough to show how the different degrees of hardness of different steels can be compared and what advantages can be derived from this comparison.

The same method also provides the means for [283] discovering how the two essential properties which are under investigation are combined in one steel. For instance, my tests may tell me that the fine grain has as much hardness, or almost as much hardness, as the mixed grain — the second type of grain. Since fine grain is in itself an advantage and since, in addition, it is combined with the advantage of body, tools made of this steel will never be given anything but the quench that results in fine grain. On the other hand, if the steel that is to be employed has a much greater hardness in its mixed grain than in its fine grain, and unless its mixed grain is extremely coarse, a tool made from this steel will be given the quench that results in mixed grain. If I recognize that the white [or coarse] grain of a steel is not much harder than the mixed grain, I shall always quench the steel at the degree of heat that will result in mixed grain, even if I need a very hard tool.

Of the three essential properties of steel, body is the one whose determination, in order to make it subject to rules, requires most in the way of preparations. I have succeeded in determining the degree of toughness, or the degree of least rigidity, left to a quenched steel. In order to decide which of two steels has more body, all that is really necessary is to be able to judge which one of two steels of the same dimensions in every direction, after both have been quenched at the same degree of heat — which one, I said — can be bent [284] more. But this also involves two difficulties. The first is to give two pieces of steel exactly the same dimensions, and the second is to heat both to the same degree. After these two difficulties have been overcome, all that remains is to have a reproducible method of bending them, one that permits to determine different degrees of bending, which is not difficult.

In order to make two pieces of steel quite alike in all dimensions, I know only one sure way, which is to have them drawn into wire. We can be satisfactorily certain of their gauge after they have been passed through the same holes several times. One could forge them in dies or else work them with the file and measure their thickness; but all this would not approach the precision of the wire-drawing die and would take much longer.

Suppose, then, that we have made wires of equal diameter of

the steels which are to be compared for body. We are still faced with the great difficulty that the entire length of both must be heated to the same degree before they are quenched. This would, indeed, be an insurmountable difficulty if it were necessary to lay them directly on the charcoal as is ordinarily done. The heat of a fire never acts uniformly upon two pieces of steel, nor even upon the entire length of one, for the pieces of charcoal are unequally distributed, and the wind of the bellows blows unequally upon the charcoal. In order [285] to heat them with perfect uniformity, both must therefore be surrounded over their entire length by an equal quantity of equally hot matter. This gave me the idea to let them get hot by immersing them in molten lead. First, I took a round (Pl. III, Fig. 12), exactly cylindrical crucible, similar to those I discussed for the trials on a small scale but of much smaller diameter, in order not to be forced to melt a larger quantity of lead than required by the experiment. The crucible was 8–9 inches long — that is, a few inches longer than the steel wire the body of which I wished to determine. In the beginning one cannot think of everything. The trouble is that, when lead is kept molten for a long time, part of this metal is lost if the crucible is made of a porous clay. The lead penetrates the walls of the crucible; it imbues them and then drips out. For ironworkers it would be easier to use the barrel of a pistol or a piece of a musket barrel (Pl. III) as a crucible, provided the barrel is well welded throughout; it must have no channels. Even if there is only the tiniest crack to permit the opening of a passage, the lead will find it, and you will see it come out in drops which may be more or less big depending on the size of the exits it has made for itself. However, no great sums of money are involved. We have time to make very many experiments of the kind I am suggesting before a [286] quarter of a pound of lead has run out of our little iron crucible.

If one wishes to use earthenware crucibles, tin may be melted in them instead of lead; it does not pass through the walls with anything like the same ease.

Regardless of whether lead is molten or tin, regardless of whether an earthenware crucible is chosen or a piece of musket barrel, we shall not be content with making the metal liquid but give it the highest degree of heat the forge is able to produce. The prepared steel wires will then be plunged into this lead bath and will be held in it. Being completely surrounded by the same, uniformly hot matter, the equal wires will take on an equal degree of heat. If they are left there for any desired time, provided they are withdrawn together and are quenched simultaneously in cold water, they will both have been quenched at the same degree.

After these wires or any other pieces of steel have been held in the molten lead or tin for a certain time, we have reason to believe that the steel has been given all the heat to which it can be brought in this metallic liquid and that this degree of heat will be a constant degree, a determinate degree of heat; that under certain circumstances, of which I shall speak later, it can be considered to be a fixed degree of heat. One of the most interesting [287] experiments of the new physical science is that performed by M. Amontons,[f] who observed that water which was brought to the boiling point had acquired the highest degree of heat to which it could be brought. It is useless to urge the fire; it cannot give it anything additional. Lead and tin which just start to melt may be compared with ice which just starts to thaw. The heat to which these metallic liquids can be brought is considerably greater than that of boiling water, but it probably is a determinate degree of heat, which has its limits as the degree of heat of boiling water has its limits. One

[f] *G. Amontons (1663–1705), French physicist who made important observations on thermometry, though in the particular observation referred to he was foreshadowed by Newton and Huygens. It is interesting to see this early evidence of Réaumur's interest in heat measurement, which culminated in his thermometer and temperature scale* (Mém. Acad. Sci., *1730, p. 452). He was wrong, of course, in supposing that in the forge fire tin or lead would reach a constant temperature through boiling but quite right in believing in the existence of such a point. Lead boils at 1740°C, tin at 2270°C, while the forge fire would not exceed about 1450°C. A lead bath is nevertheless a very useful means of quickly and uniformly heating parts to be quenched and prevents scaling or decarburization.*

produces this degree of heat in lead or tin when the fire has been driven as violently as possible and perhaps much earlier. What I just said of lead and tin probably is also true of all metallic liquids. Apparently there is a limit beyond which, when it has once been reached, ordinary fire adds nothing more to their heat, and this limit is most likely different for different metals. It would be interesting to know this limit and to know what relationship exists among the highest degrees of heat of which the metals, all the liquids, and other substances are susceptible and what relationship exists between these highest degrees of heat and the specific weights of these substances and also between them and the difficulty we experience in melting them. There was a time when I made a great many experiments in this field, but I [288] have made not nearly enough to have come to any conclusion.

One thing seems very certain — that other metals are susceptible of a higher degree of heat [than lead and tin] and that each has a different *maximum* heat. To prove this, melt lead in a crucible; melt the same amount of pig iron in another similar crucible, increase the degree of heat of these two metallic liquids as far as it is possible, and plunge into both an equal piece of iron. The end of the piece that was plunged into the iron bath will take on the whitest color that fire can produce in iron; and, if nothing heavier than an iron wire was used, the iron wire will melt in the bath. But the iron wire that was immersed in the lead (or tin) will be far from having melted.

As we use lead in order to heat equally hot any steels which we wish to test for body, so shall we use molten iron to heat equally hot any steels whose range of different grain sizes we wish to observe accurately. The iron can be put in ordinary crucibles; it is not necessary to use an iron crucible, as for lead. This, then, constitutes the reliable method I promised for heating equally any steels whose different types of grain are to be compared.

Let us come back once more to the testing of the body of steel. [289] When wires of the same thickness and length are plunged into molten lead or tin or even in an iron bath, if desired, and are simultaneously withdrawn and then quenched, they have been quenched as identically as possible. If one of the two is more flexible than the other, it has more body. There are many possible methods of measuring their flexibility. The one I have most frequently used requires the construction of a small machine[g] (Pl. IV). It consists of a copper or iron plate (Fig. 1, *BB*) placed horizontally and held by screws on a piece of iron that is bent twice into a right-angled brace. I shall not give the dimensions of the different pieces of this machine, for they are rather arbitrary, and with the scale given [at the bottom] on Plate IV the dimensions of the machine that I had made can be found. At one side of the plate there is a little vise (Fig. 1, *C*) which is attached to the same piece that supports this plate. The purpose of this vise is to hold one end of the steel wire the body of which is to be determined. The two arms of the rectangular brace which supports the plate extend beyond the plate. Either by bending them at right angles near their ends or by adding another small piece, each of these ends is made to be higher than the rest and is provided with a square opening (Fig. 1, *DD*); it has something like a lunette guide, similar to that of some lathes. The two guides, or openings, receive a piece of iron forged into a billet (Fig. 1, *E*) which is longer than the plate. It can be moved to and fro [290] in the guides. This piece is thicker in the middle, where it is traversed by a threaded hole into which a long, coarse screw (*F G*) is fitted. One of the arms which grasp the piece holding the coarse screw is also pierced from above by a screw hole into which a small screw (*I*) fits, which, when tightened,

---

[g] *This is the first materials-testing machine on record, although Leonardo da Vinci, Galileo, and Hooke had all performed simple experiments to determine the breaking strengths of wires by hanging weights on them until fracture resulted. Shortly after the publication of Réaumur's work, Musschenbroek in Holland published an excellent study of the mechanical properties of materials of all kinds* (Introductio ad cohaerentium corporum firmorum [Leyden, 1729]). *His machine was a greatly improved version of one used by Desaguliers* (Phil. Trans. Royal Soc., XXXIII [1724], 345–47).

holds this piece firmly in place. As this piece can be moved in the guides whenever desired, it is obvious that the coarse screw can be held at different distances from the vise.

Using this machine is simple. One end of the quenched iron wire (NO) is clamped in the vise. In doing so, one is careful to place the wire horizontally and so that it has the same elevation as the screw. After the screw has been turned until it touches the wire, it will push the wire forward when turned further. It will force the wire to bend, and, finally, after the screw has advanced to a certain point, the flexibility of the wire will have been strained to the limit, and the wire will break.

The location on the plate up to which the screw has advanced when the wire breaks will be noted down. In order to make this easier, the plate can be divided into as many parts as desired by drawing lines perpendicularly to the direction of the screw. Then take the broken steel wire from the vise, put another one in its place, let the screw gradually force it to bend more and more, and observe once more where the screw was when the wire broke. The more flexible of these two wires, the one that has [291] more body, is the one that before breaking permitted the screw to advance farther. The difference between the two locations gives a measure of the flexibility, or the body, of the steel.

I should mention that the end of the coarse screw (G) must be drilled hollow like that of a key and that there must be a little piece with a round spindle which can be put into this hole (H) and turn in it freely. Outside the screw, this spindle must end in a small fork in which the iron wire is placed. Otherwise it would be difficult for the screw to push the wire ahead, as the wire would slide over or under it.

Instead of graduating the plate, it would be still easier to have a small graduated rule, which can be moved over the plate. A piece of iron which is bent so that its two parts are parallel to each other can be used as such a rule (Fig. 1, KL). The distance between the two parts will be only a little greater than the thickness of the plate. The plate fits easily into the space between these two parts. The lower one will have a screw (Fig. 2, *M*) which, when tightened, will hold the rule in the place selected for it. The upper part, which is the rule proper, will have divisions at intervals of a line or at closer intervals if desired.

This is how the little rule is used: When the iron that is to be tested is in place — that means when its one end has been clamped between the jaws of the vise and the other [292] end has been placed in the small fork of the coarse screw — the rule will be brought over and will be made fast, so that its end is quite close to the end of the lower tine of the little fork. The coarse screw is then turned slowly, and, as we turn it, we observe with which division on the rule the end of the lower branch of the little fork is even. The moment the wire breaks, we stop turning the screw and note down the division of the rule which one end of the fork had reached when the wire broke. For instance, if it had arrived at the eighth line, at the eighth division, and if when some other wire is broken the end of the same line advances only to the sixth division, we know that the first wire is made of steel with more body.

The wires that are placed in the machine will not always be as straight as one might wish them to be; they often warp in quenching. The shorter they are, the less they will be inclined to warp, but they will also bend less before breaking. However, it rarely happens that they warp enough to make it impossible to judge the outcome of the experiment.

After two steels have been compared concerning their body, it should not be neglected also to compare their grain; for, if the steel with more body is found at the same time to have a coarser grain, it indicates that it would have much more body if it were quenched less hot in order to give it the type of grain [293] of the other one. One would reason in the opposite way if the steel that broke closer to the end of the rule had the coarser grain; by quenching it less hot, it could be given more body and the type of grain of the first one.

Body, or flexibility, could also be determined by means of weights. It is enough to make this suggestion without supplying sketches or even an explanation of the way in which a cord could be fastened to the steel, and how this cord would then pass over a small pulley and would be loaded at its other end with weights, to which others would be added as long as it is necessary.[h]

For the makers of springs for watches and clocks it is more important than for any other artisan to obtain steels that have body. They treat their steels by tempering, an operation of which it is not yet time to speak. They must coil their steels; but those steels which have no body cannot be coiled. We could determine the body of steels by a method that derives from the way they are processed — I mean by using small, hollow wheels of different diameters, something like rings (Pl. III, Figs. 14, 15). The steel that could be wound upon the ring of the smallest diameter would be the steel with the most body; but then it would be necessary to forge the steel into thin strip or to draw it into fine wire, because otherwise it would not have enough flexibility to withstand the test. One end of the strip or wire would be fastened on the ring by a screw (Fig. 15, *a*), and the artisan would then attempt to fit the rest [294] of the steel on to the circumference of the ring, either by hand or with a small hook (*b d*).

These methods of testing the body of steel really are as precise as one could wish; but they require a little too much in the way of apparatus to be adopted as ordinary practice. I doubt that we shall become as progressive as these methods show we could become. At least, we could hardly expect the ordinary artisan to have recourse to them. Those who use the steel are far from being

---

[h] *This second method would, of course, measure the bending strength as well as the maximum strain in bending which Réaumur had earlier defined as body. The term "body" is currently used to imply a vague aggregate quality, a combination of hardenability, toughness, and reliability in service. Réaumur is far ahead of his time in attempting to reduce it to quantitative measure and shows wisdom in matching the type of test to the intended service.*

equipped with the machines needed to draw it into wire, and, although this operation is very simple, they might get into difficulties. They would need a drawplate and something corresponding to the *bench* of the goldsmith. They also could use a small jack similar to the ones described by M. Dalesme in the *Memoirs* of the Academy of 1717.[1]

Instead of these precise methods, a cruder one could be used which is still good and provides a means to determine the hardness of a steel at the same time as its body is being tested. This is practically the only test known to our artisans, and I shall suggest some improvements which will be useful. In order to try out a steel, a cold chisel is forged from it. Either the chisel is forged entirely of steel or only its end is made of steel, which is sufficient. The chisel will be quenched, and the degree of heat, the color at which it has been quenched, will be noted down. Then one [295] tries out if this chisel can cut iron cold. If the edge folds back, it will be heated and quenched hotter, for the first quench has not given it the necessary hardness. If it spalls after the second quench, it has not been left enough body. One will then aim at quenching it at a degree of heat intermediate between the two previously used. If none can be found at which this steel does not fail because of one of the two extreme possibilities — that is, either by folding back or by spalling — this is a sure indication that this steel cannot combine sufficient body with much hardness. Instead of being quenched at lower degrees of heat, it is additionally tempered; but this is not yet the time to talk about tempering.

If the chisel successfully cuts iron cold, it can be made to cut it in two different ways, either obliquely or perpendicularly (Pl. III, Fig. 17). If it is used to cut obliquely (*m*), it is usually made to rest on the long edge of the bar that is to be cut. Chips are taken off as they would be taken off a piece of wood, first thin ones and then thicker ones, which means that the chisel is held sometimes more

---

[1] Dalesme, "Crics nouveaux," Mém. Acad. Sci., *1717, pp. 301–4.*

and sometimes less inclined and that it is more or less strongly hammered, depending on the size of the chip that is to be cut off. The cut in the iron, the spot from which the chip has been taken (*n*), itself gives an indication of the hardness of the steel. The harder a steel is, the cleaner, more sparkling, and brilliant will be the cut. A steel is given a good rating if it lifts [296] large chips without spalling or breaking. But, if the steel spalls, the rating it should be given is not so certain. If the hand holding the chisel wavers when the chisel is struck with the hammer, the edge will break without fail, even if it has sufficient body.

This kind of test provides no means of comparison; it is not able to tell whether one steel outranks another; but it can be made more precise as follows: Out of the worst grade of steel, the one of the least hardness (Pl. III, Fig. 16), I have a thin bar forged. This is quenched mildly at one end (*f*); I then divide it into several parts, like a rule, and use it as a measure of hardness and body of steel. After a chisel has been forged out of steel, I have it successively quenched and tempered at different degrees of heat, and I also give its cutting edge different degrees of thickness. I then determine after each operation up to which location, up to which division and which part of this division, my chisel can cut the quenched steel without spalling. When I have found the limit beyond which I cannot make it cut into the steel, I note it down. Let us suppose that our chisel could make no impression beyond the third line. I have a similar chisel forged out of another steel and determine which kind of quenching and tempering gives it the greatest hardness and body. If I subsequently succeed in making this second chisel cut up to about the second division, I know that this steel has more body than the other one [297] and probably more hardness.

There are few chisels made of steel which can cut another quenched steel, even where its grain is fine. In order to be able to test all classes of steel, they must be opposed to something less hard. For this purpose, I am using a flat bar of a naturally hard iron, one end of which has been heated and quenched almost melting-hot. Not every steel can cut into this end, not even at some distance from the actual end. Thus I could graduate, and I have graduated, such an iron bar for the purpose of comparing the quality of different steels.

Instead of using a hammer, one could strike the chisel, placed vertically, by means of a weight, which would fall from a certain constant height in a constant direction. This could be done easily by employing a machine similar to the one with which pinmakers form the heads of pins, and to my mind this would be a good method of selecting the steels that are most suitable for very sharp cutting tools. I assume that I have a number of chisels made of different steels that are to be tested; they have all been given the same width and approximately the same angle, and they have been used to cut iron or wood or some other material. It is clear that the best-cutting steel would be the one that, struck less often, would cut a piece of equal or greater thickness. If one chisel cuts only after thirty blows what another cuts [298] after twenty, one has a means of judging the difference in quality between the steels from which these chisels were made. But at the present state of the arts, as long as we do not attempt to encourage high-minded competition among our artisans, as long as we neglect to reward those who distinguish themselves in their profession, we cannot expect that they will devote their time to similar investigations and that they will do anything with precision.

Besides veins of iron which may be found in steel, other irregularities may be encountered in their texture which are not so easily recognized. We have seen that these textures vary from each other in their degrees of fineness and hardness. These differences, which we have considered only in different steels, can be found in the texture of one and the same steel and can make it defective. The best method known to me for recognizing if a steel is of the same quality throughout is to have a piece of it worked on a lathe, to have a cylinder made of it, which may be more or less long, more or less big, as desired.[j] Watchmakers must often have such cylinders turned, as they use them for the shafts of wheels and gears, etc. They know from experience that there are steels with a non-uniform texture; the graver cuts deeper into spots of less hardness, digs into them more deeply than into the rest. Whereas our other tests are meant for quenched steels, this one is used on unquenched steels.

In general, the [299] quality of unquenched steels can also be judged to some extent by examining their fractures. I should like to see the careful artisan equipped with the fractures of all grades of unquenched steels as well as of quenched steels. The thing that is shown most plainly by the fractures of unquenched steel is whether a steel contains grains or platelets of iron; for, provided they are present, they show up by a brilliance which permits to distinguish them from the other grains. If there is a scattering of them in quenched steel, these platelets of iron are not so easily recognized. It would be impossible to distinguish them among the [coarse] white grains. Among the mixed grains they may be mistaken for part of the mixture. And, if they are present among the fine grains, one may sometimes wonder if they are not still part of the mixed grain.

Another thing I have observed in the fractures of unquenched and very well-annealed steels is that, as the grains of these fractures become finer, they become grayer, and the steels become more difficult to work; those with coarser grains are more easy to use. It is the same with those of a less dull color.

I believed that I could compare the hardness of two different steels after they had been forged in a swage, in which they were given the shape of prisms with the same right-angled triangle as

[j] *One of the mysteries of metallurgical history is why etching was so late in being used to distinguish between metals and to show heterogeneity. Etching was used for decorative purposes on armor and weapons very early, both through a ground with a scratched design and to show the water in Damascus blades. Perret (1771) used nitric acid to distinguish between iron and steel, but it was not until the nineteenth century that etching was used to show structure.*

bases. It seemed to me that, if the right angle of [300] the one prism were placed upon the right angle of the other one (Pl. IV, Fig. 3), and the upper one were then struck with a hammer, the harder of the two steels ought to notch the other one. However, I usually found both of them notched. The blow is not as instantaneous as it would seem. The upper prism is propelled forward. If, at first, it has forced the parts of the lower prism to yield, or to break, the sharp edge of the lower angle will have become thicker after it has yielded, or has broken, and is then, in turn, able to force the sharp edge of the upper angle to yield or to break. The same would have happened if the upper angle had yielded first. I am mentioning this experiment only so that those who might have the same idea as I had will be saved the trouble of trying it out.

The comments accompanying the Plates, which were purposely made very detailed, should help to clarify anything that might still be obscure about the means I have suggested for recognizing the different qualities of steel and for comparing those of different steels.

EXPLANATION OF PLATE II

Figure 1 is a billet of intractable steel. Its surface is completely covered with cracks.

Figure 2 is another steel billet with cracks but with fewer cracks than the preceding one. Here they occur principally at the corners, as in A. The cracks are sometimes so fine that one has to look at them closely to discover them.

Part DB of the piece of steel shown in Figure 3 has been bent around to make it approach DC. This steel, being of bad quality and hardly fit to be worked, partly broke in the bend D, although it was bent hot.

Figures 4–7 are fractures of different bars of steel, each one of which has one of the spots called "roses." The rose of each bar differs from the others in color and shape.

Figure 8 is a small steel bar; the end F has been hammered out into a point. It then has been heated and quenched. If the thick end L is struck with a hammer, the point $(F\ G)$ will break. In the same figure, M marks the defects called "laps," which are parts badly welded to the rest and which sometimes, when they are thin, can be lifted off.

Figure 9 consists of four separate pieces of steel, fg, gh, hi, and ik. They were obtained by striking the bar of Figure 8, which [302] successively broke at G, H, I, and K. It will be assumed that the point was almost melting-hot, while the end L remained almost cold or did not become colored at all. Consequently, the fractures near F have a coarser grain than those farther down, a fact which can be observed in gh, hi, and ik.

Figure 10 shows a flat bar of steel which was forge-welded to a bar of iron so that afterward the steel could be broken in such a way that the fracture would show the entire range of different grain sizes.[k]

Figure 11 is the iron bar.

Figure 12 shows how these bars are partly welded together. They are completely welded at aa and down to b; dba is the iron bar, cha is the steel bar, and de is the line along which the iron bar must be cut after it has been welded.

In Figure 13 the piece is ready; $ghh\ iik$ is the groove cut into the iron. The part beyond the groove is steel. All that is still to be done is to heat one end of this piece melting-hot and then quench it. Thereafter it can be broken lengthwise.

Figure 14 shows the piece of the preceding figure, which has been partly cut at $l\ l$. With the chisel n, it is being forced to open up at $m\ m$ and the surrounding area.

Figure 15 illustrates one of the two halves obtained by the division of the previous piece. Surface $pqss$ shows where the iron was cut with the chisel or the file which made the groove; $opqr$ is the fracture of the steel, on which

---

[k] *Harry Brearley* (Heat Treatment of Tool Steel [*London, 1916*], *p. 91*) *refers to this test — modified only by milling a notch instead of welding on iron — as "Alling's method" and shows photographs of such fractured pieces which illustrate more clearly than Réaumur's engravings the sequence of fractures involved.*

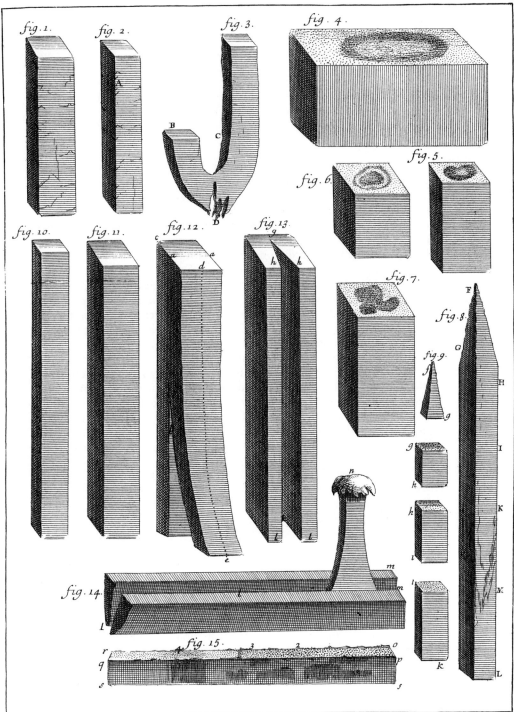

Plate II

all the different types of grain can be seen; *o, 1* is the first type, the coarse grain. *1, 2* is the second type, or mixed grain; *2, 3* is the third type, the fine grain; *3, 4*, and everything thereafter is the grain of the grade of steel under consideration when it has not been affected by quenching; it is the fourth type. Part *o u* consists of grains of less whiteness and brilliance than those which follow within the same division, the reason being that the end was heated so hot that it was almost melting.

EXPLANATION OF PLATE III

Figure 1 shows a steel bar forged razor-shaped, so that it can easily be broken lengthwise. *AA* is the back, which is much thicker than the rest. *BC* is a groove cut with the chisel along a line where this bar is to be broken. If end *C* is the one that is not supposed to be affected by the quench, no groove is cut near *C*.

Figure 2 shows the preceding bar, which now has been broken all along the groove, *BC*. The fracture, *DE* shows the different types of grain. This is a coarse steel of very bad quality. Its coarse grain is neither so white nor so sparkling as that of some other steels. But it is even more noteworthy that these grains seem to be arranged in parallel layers, a fact which indicates the presence of iron. These layers are still more noticeable in the mixed-grain division (*1, 2*), where something like fibers can be seen, which are pure iron. These fibers are still more evident in the fine grain (*2, 3*). Here we have another rule — that the presence of iron in steel is always more noticeable in the fine grain than in the preceding divisions. In the first division it exists as white grains that can hardly be distinguished from those of steel, except when they are tried with the file. The fracture of the fourth type of grain (*3, 4*) resembles that of fibrous irons.

Figure 3 consists of four pieces (*F F F F*) which have been separated from the bar of Figure 1.

Figure 4 is a bar of another steel. It is seen by its fracture that it is still coarse but less so than the previous steel. A small vein of iron (*i*) appears in the mixed grain (*1, 2*) and two larger ones in the fine grain (*2, 3*). The grains of the third division in this steel, and of the preceding one, are not as fine as those of the next figure. The fourth type (*3, 4*) consists partly of grains and partly of fibers.

Figure 5 shows (as the earlier ones) the fracture, *LM*, of a steel which is finer than the preceding ones and actually one of the finest kind. The coarse grain is in no way mixed, the mixed grain (*1, 2*) is very mixed, and the fine grain (*2, 3*) consists of very fine grains, which are well separated from each other. The unquenched part has only grains, which are very fine, so that this division actually equals the third division in some quenched steels, but the grains are grayer than the fine grains of this same steel. Such very fine grain at a location where the steel was not affected by the quench ordinarily indicates a steel which is hard to process and which has little body.

Figures 6 and 7 show the fractures of different steels. It was considered sufficient in this case to draw the surfaces of these fractures. It is assumed that the two steel bars on which the fractures have been drawn were heated equally by being plunged up to *R R* into a bath of molten iron. *O, 1* is the first division of the one steel and *Q 1 M* of the other; *3* indicates the end of the fine grain in both steels. It will be noticed, and it is the purpose of these

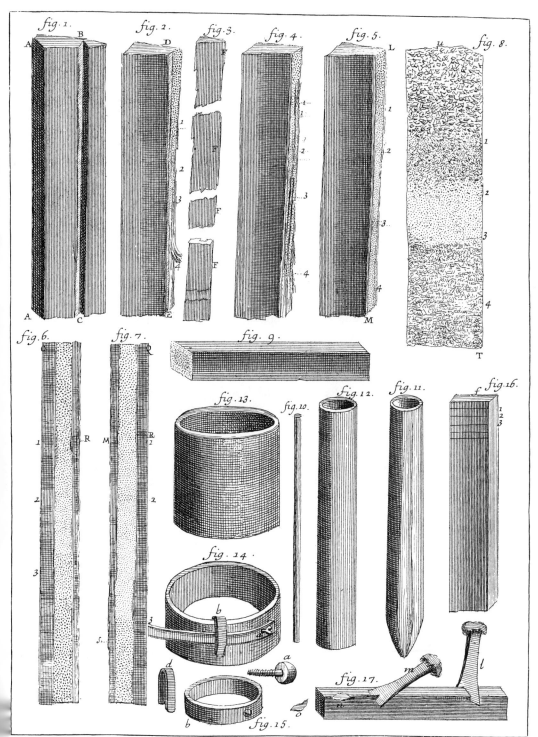

Plate III

figures to point out, that the steel of Figure 7, which is the finer of the two, has a longer range of [quenched] grain (Q, 3) than the steel of Figure 6 (O, 3). It will also be noticed that in Figure 7 the ratio of the extent of fine grain (2, 3) to the extent of mixed grain (2, 1) is greater than the corresponding ratio in Figure 6. The unquenched grain (s) in Figure 7 consists of finer grains than the corresponding division in Figure 6.

Figure 8 shows the entire range of different types of grain of a steel as seen enlarged under a magnifying glass. The figure had to be short, because it is assumed that it was obtained on a piece of steel only a short end of which was quenched. *u*, *1* is the first division, which seems to consist of platelets similar to those observed in certain irons by the unaided eye. This demonstrates that steel differs from iron less by the shape than by the smallness of its parts. 1, 2 is the second division, which has brilliant platelets mixed with dull grains. In the third division (2, 3) there seems to be nothing but some sort of grain. It should not be concluded, however, that the particles have not here, as elsewhere, the shape of platelets. The only conclusion that can be drawn is that the microscope, which in the first and second divisions shows platelets of the size indicated, does not show them in this one; but platelets would be seen under a much stronger microscope. 3, 4, *T* is the structure unaffected by quenching.

Figure 9 shows a steel bar the end of which was broken. This steel was treated [i.e., decarburized] as described in the eighth memoir. The circumference of the fracture is a layer of iron, and all the rest is steel.

Figure 10 shows a steel wire which is to be tested for body.

Figure 11 is the barrel of a small pistol, or a piece of a musket barrel, in which lead or tin can be melted, so that steel wires can be heated equally when their body is to be compared.

Figures 12 and 13 are crucibles which can be used for the same purpose or to keep iron molten, in which steel bars (forged into razors like those shown in preceding figures) can be heated equally to compare their body.

Figure 14 shows a ring, or hollow cylinder, upon which steel strip or steel wire can be wound to test its body. *a* is the screw that holds one end of the iron [*sic*] strip or wire; *b* is a hook which keeps the steel fastened to the ring; *d b* is the hook viewed separately.

Figure 15 shows another ring of smaller diameter. Experience will indicate what diameters should be given to different rings to make the tests.

Figure 16 is a bar of bad steel, or of hard iron, the end (*f*) of which has been quenched. It is graduated by parallel lines (1, 2, 3). It serves to test chisels made of different steels. The best ones are those that cut closest to *f* without spalling and without having their edges fold back.

Figure 17 shows a piece of iron with two chisels resting on it; *l* is a chisel standing upright; *m* is a chisel placed obliquely; *n* is a cut which was previously made with the chisel held obliquely. The brighter and cleaner this cut looks, the better is the steel which has made it; *o* is a chip that has been lifted by the chisel.

EXPLANATION OF PLATE IV

Figure 1 is the little machine for testing the body of steel, as viewed from above. It can be held in one's hand, but it is much more convenient to fasten it in a vise.

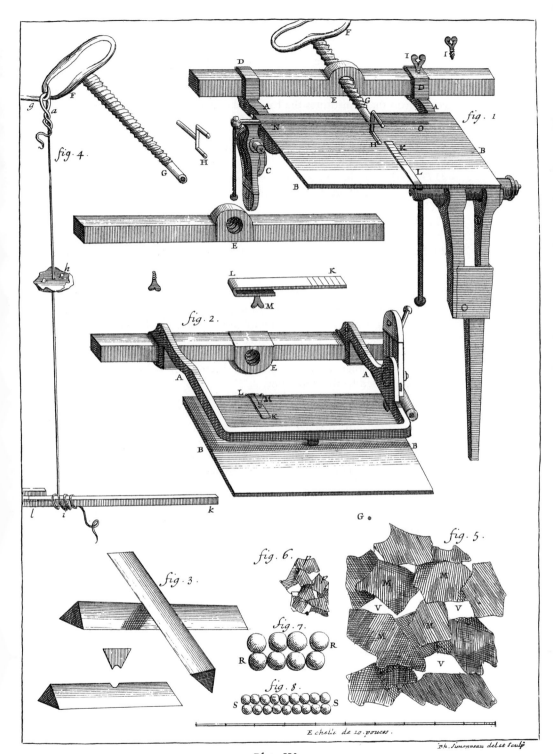

Plate IV

Figure 2 shows the same machine, as viewed from below.

Some parts of this machine are also shown separately, but the same letters are used throughout for the same parts. *A A* are the two arms of the bent piece of iron that constitutes the base of the machine.

B B is the plate, made of iron or copper, which is attached to these two arms.

C is the small vise, the jaws of which hold one end of the wire that is to be tested for body.

D D are the two lunette guides, or holes, that receive the movable shaft.

E is this shaft, provided with a screw hole at *E*.

F G is the coarse screw.

H is the small fork which fits into the end of the coarse screw and turns in it freely.

I is a small screw which serves to secure the shaft which holds the coarse screw. This screw is placed closer to the vise or farther away from it, depending on the length of the wire that is to be tested.

K L, in Figure 1, is the small rule, which is bent at *L*.

L M, in Figure 2, is the same rule. *M* is the screw that holds it to the plate.

N O, in Figure 1, is the iron [sic] wire in position to be broken. It will be broken when the screw *F G* has strained it to a certain extent.

Figure 3 shows two steel prisms which are equal in every respect, one placed upon the other. Both of them have been notched by the same blow which was applied to the upper one.

Figures 4 ff. explain matters discussed in the eleventh memoir. *g h i* is a steel wire; its upper end is hung up at *g*; *h* is a small plate on which one or two pieces of red-hot charcoal are placed to heat the wire at this location to the point that is required to quench it; *k l* is a lever around which the lower end of the steel wire is wrapped. The arm *i k* is much longer than the other one; it is made as long as one finds it necessary. The arm *i l* is stopped at *l* by any suitable body. The arm *k i* is loaded to break the steel wire.

Figure 5 shows a grain of steel as it would look if it were vastly enlarged. Its natural size is shown in *G*. *M M M* are the molecules of which this grain is composed. *V V* are the voids left between them.

Figure 6 shows part of this grain, or one molecule, by itself; *p p* are the parts of which this molecule is composed.

Figure 7 consists of two rows (*R R*) of equal spheres, some of which do not touch.

Figure 8 consists of two rows (*S S*) of smaller spheres than the preceding ones, all of which touch.

CHAPTER SIX

The Preparation of Steel out of Iron by Cementation and by Fusion

Johann Andreas Cramer

Extract from *Elementa Artis Docimasticae* (Leyden, 1739). Reproduced from the English translation by Cromwell Mortimer (London, 1741), pp. 344–50.

Editor's Note

The most "scientific" parts of the great sixteenth-century metallurgical books were the sections describing the quantitative techniques that were in use in the well-equipped assay laboratories of the time. Even in the best of these, no attempt whatever was made to provide theoretical justification for the operations. The chemical reactions discovered and used by the assayers, however, greatly influenced theoretically minded chemists and provided the basis for much of the early speculations on chemical affinity. As time advanced, the assayers themselves began to feel the need for some theory. The first book on assaying in which the well-known processes are put into something like a scientific framework was the Elementa Artis Docimasticae *written by Johann Andreas Cramer (1710–1777) and published in Leyden in 1739. Cramer, subdividing his book into a theoretical and a practical part, attempted a logical arrangement of all the different kinds of separatory processes used with metals and minerals. Like most leading chemists at the time, he was strongly influenced by the phlogiston theory as expounded in 1703 by G. E. Stahl, a metallurgical chemist. The presence of a nearly intangible reducing principle had been suggested earlier by J. J. Becher in order to account for the observation that those substances that had the property of converting ores to metal usually also possessed combustibility. Phlogiston was a descendant of the salt-sulphur-mercury theory[a] of Paracelsus (1493–1541): in fact, many eighteenth-century chemists continued to use the*

---

[a] *The significance of the choice of the three archetypes should be noted, for they represent three of the four types of bonding in today's quantum-mechanical theory of*

term "sulphur" for the reducing principle with no thought whatever of the substantial elemental material that was then as now known by that name.

It was inevitable that steel, which was produced by prolonged heating of iron in a phlogiston-rich environment, should come to be regarded simply as iron freed from earthy parts but containing added phlogiston and therefore in the state most nearly approaching that of fullest metallicity — a conclusion quite in line with the early Aristotelian view. Cramer's two chapters on steel, which combine the phlogiston theory with clear experimental instructions, follow, reproduced directly from the 1741 English edition.

---

solids. In the development of science, the unraveling of composition by the analytical chemist and the discovery of the electron (which is not far from being phlogiston itself) by the physicist had to occur before there could be useful scientific study of the changes of properties of matter. Nevertheless it was the apparent changes in physical properties of substances that suggested the idea of transmutation, and in a physical sense, if not chemically, the alchemists' goal was a real and a valid one. (See C. S. Smith, "The Prehistory of Solid-State Physics," Physics Today, 1965, 18 (No. 12), 18–30.)

The Preparation of Steel out of Iron, by Cementation.

STEEL is made of Iron in two different Manners, *viz.* by Cementation (*Part* I. § 459.), or by Fusion. The Cementation is performed in the following Manner.

APPARATUS.

1. Chuse some Barrs of pure Iron, not over-thick, and quite free from heterogeneous Matters, the Flexibleness of it, both when hot and when cold, is a very good Sign thereof. Prepare a Cement composed of such Ingredients as emit an abundant Phlogiston, when agitated by the Fire, provided the said Phlogiston be altogether free from the sulphureous mineral Acid: Such as are all extinguished Coals, and in short all Parts of Animals and Vegetables; among which, those, however, are much so, which contain a greater Quantity of Oil in them, and which being freed of an excessive Phlegm, have been burnt before into a semi-carbonaceous Mass. Avoid whatever absorbs oily Vapous with great Force, or even spreads the Acid of Sulphur, or the mineral Sulphur itself. It is better to add a few Compositions, in Order to clear the Matter.

Take Charcoal-dust moderately pulverised 1 P^t. of Wood-ashes $\frac{1}{2}$ P^t. Mix them together.

Take Charcoal-dust 2 P^{ts}. Bones, Horns, Leather, Hairs of Animals (it is all one if you use but one, or several, or even all of them mixt together; for one of them alone is as sufficient as the Mixture of them all) burnt with a gentle Fire till they are black in a close Vessel, then pulverise them 1 P^t. Wood-ashes $\frac{1}{2}$ P^t. Mix them together.

As for the reft, it has been found, that the Parts of Animals, on Account of the Abundance of Oil they contain, are of a quicker Effect than the Reft.

2. Prepare an Earthen-veffel, the beft Figure of which is the Cylindrical, two or three Inches higher than the Iron-barrs N°. 1.) are long: Put into the Bottom of it your Cement prepared in the aforefaid Manner, fo that being gently preffed down, it may cover the Bottom of the Veffel to the Height of one Inch and a half. Place the Iron-bars perpendicularly, in fuch Manner, that they may be about one Inch diftant from the Sides of the Veffel, and from each other: Fill the empty Interftices with the fame Cement; and cover alfo the Bars with it, that the Veffel may be quite full; next cover it with a Tile, and ftop the Joints with thin Lute.

3. When thus prepared (N°. 2.) put this Veffel in a Furnace, where you may for feveral Hours maintain an equal Fire, as either in the Bottom of the Tower, or in the firft Chamber of the Athanor (*Part* I. § 243.) Make a Fire fo ftrong, as that the Veffel may be moderately red-hot for fix or ten Hours together: When this Time is over, take it out of the Fire, and dip the red Iron-barrs into cold Water. They will then be brittle, and turned to Steel, there will appear no Scoria at the Out-fide, nor will the Weight be diminifhed, if you have but rightly made your Procefs according to the Regimen of the Fire.

4. The Signs of the Iron's being changed into Steel are, if being red-hot, and extinguifhed in cold Water, it becomes very hard, not yielding to the Hammer, brittle when more ftrongly hammered, and refifting the hardeft File: By which Quality it is diftinguifhed from Iron rendered malleable, which indeed grows rigid when extinguifhed in Water, but yet retains a confiderable Degree of Ductility in the Cold, and may be extended in all Dimenfions with the Hammer. However, Steel that is cooled foftly, and by flow Degrees, may be filed and extended with

the

the Hammer any Way, some more, some less: By which Quality it may be distinguished from crude melted Iron: For this is often brittle, both when cold and when hot, though it has not been extinguished in Water. But, there are a vast many Degrees in the hardening of Steel: For, if it has been made too red-hot, and is suddenly extinguished in cold Water in Motion, it hardens more than if it had been but faintly red, and cooled in warm Water. This Hardening is caused by all such Bodies as suddenly absorb the Heat, and at the same Time do not easily penetrate the Steel, but change its Nature. Steel is moreover of a darker Colour, and the Surface of it, when broken, appears to consist of smaller granulated, and even striated Particles, than the Iron which it is made of: The Germans call it * Klar-koznig, Klar-speissig. But, this appears more distinctly, when Steel is welded to the same Kind of Iron, which it was made of, and when the Mass made red-hot is well incorporated together, by hammering: If then you harden it again by extinguishing it in cold Water, and polish it, the Veins of Iron may be very well distinguished from those of Steel: For, the Iron-ones are more whitish, and almost of a Silver-colour, but the Steel-ones of a darker Dye, and almost of the Colour of Water. For which Reason Dr. *Stahl* is of Opinion, that the Steel of *Damascus*, which has the same Colour on the Out-side, is made in the same Manner. But, if such Steel mixt with Iron is broken, you may likewise observe the Difference of the Largeness and Colour of its Particles.

The Use and Reasons of the Process.

1. All you do in this Operation is, to apply oily Vapours to pure Iron, the rigid Body of which being mollifyed by the Heat, and made quite red hot, is penetrated by the said Vapours, which then strictly

* Fine grained, fine fibred.

unite

unite to it: Which is thought to be so, because the Iron thus changed, not only preserves its first Weight (whereas when made red hot, it otherwise loses always a great Quantity of its Substance, which goes away in Form of scaly Scoria's) but even proves to have increased it a small Matter, unless too great and long-lasting a Fire has burnt the Surface of it; which the Scales going off from it do shew. For this Reason, the essential Difference between pure Iron and Steel, consists in the greater Proportion of Phlogiston more intimately joined to one than to the other. Thence the Reason is likewise clear, why a too thick Piece of Iron being put into such a Cement, or the Iron-barrs not being left long enough in the Fire with the Cement, they are only surrounded with a steely Crust, while the inward Substance remains Iron.

2. That every oily Substance free from the Acid of Sulphur, is fit for changing Iron into Steel, is plain from the several Experiments of Workmen, some of which use for their Cements a Multitude of different Particles of the animal and vegetable Kingdoms, and yet all of them produce the very same Kind of Steel, provided the other Ingredients are alike: But if you employ for your Cement any Body exhaling Acid of Sulphur, or even Sulphur itself in a strong Fire, you not only will have no Steel, but instead of it the Substance of the Iron changes, and goes away into a Scoria. For this Reason Sea-coals * are not fit to render Iron malleable, nor to turn it into Steel: Nay, Iron and Steel are more easily burnt and destroyed by them, than they are by an open Fire

* Our Sea-coals or *Newcastle* Coals, or in general all the fossil Coals which cake in burning and run into Cinders, abound with Sulphur, and therefore are improper to be used about Iron, always making it brittle; but Pit-coals, Kennel-coals, and Scotch-coals, which burn to a White-ash like Wood, and abound more in a Bitumen, may be used in the first fluxing of the Iron from the Ore, and if the Iron prove not so malleable as is required, this Property may be given to it by melting the Metal a second Time with Wood.

of Wood-coal, unless you use a peculiar Remedy for it.

3. Therefore, the best Steel made red hot a long Time, or frequently, especially in an open Fire, the phlogistick Part being dissipated, turns to Iron again, provided the Fire is managed, so as that it may not quickly turn the whole Mass into Scoria's.

PROCESS LVII.

The Production of Steel out of crude unmalleable Iron, or out of its Ore, by Fusion.

WE shall here in a clear and general Manner give the Method for making Steel by Fusion. Chuse for Instance, Iron-Ore, or the Iron itself still crude of the first Fusion, which we know can be rendered tough and firm by being melted, made red hot, and hammered. For, according as Iron, or its Ore is different in its Kind, so you may make with it different Sorts of Steel, and with greater or less Ease, or Difficulty. Put at one or several Times into a Bed made with Charcoal Dust in a Smith's Forge, such Quantity of this Metal divided into small Parcels, as that the Metal remaining after the Melting of it is compleatly performed, may not be more than two or three common Centners; not only that the Melting may be sooner finished, but also because a small Mass may be better and more equally penetrated with the Vapours of the Phlogiston: Nay, they also add, as a defensive Menstruum, some of the vitrescent fusible Scoria's, either of Sand, or of small Stones of the same Nature: Then put upon them Abundance of Charcoal, light them, and admit only a gentle Blast of the Bellows; that the Scoria's and the Metal may both melt very well: Take out now and then Part of the Scoria's, and often stirr the melted Mass with a Stick; that all the Parts of it may as much as possible feel the same Degree of Fire. Having at last removed the Fire and the Scoria's,

ria's, and the Mass being grown solid, put it upon an Anvil, and with a Hammer divide it into two Parts, which must be extended into long Barrs by being made red hot, and hammered several Times over: Then extinguish them in cold Water, whereby they are rendered so very hard, that they will fly asunder when struck with great Force, and will not be filable; which shews that the Operation has been well made. But, if you have a Mind at the first Time to change Iron Ore or crude Iron into Steel, nothing but a Repetition of Trials will inform you, how long and at how many Times the Matter in Hand must be melted, made red hot, and hammered. For, there are Ores, which, by a first Fusion, produce Masses which are intermediate between malleable Iron and Steel; or resemble Steel that is but half-worked: Whence such Steel-Ores are commonly called by the Germans * Stahl-Stein. Other Ores, on the contrary, must often go through a Number of long-lasting Fusions and Hammerings, and sometimes lose half the Weight of the crude Iron in the first Fusion, and yet never yield a right Kind of Steel. But, it is easy to guess at the Reason, why this Process is very much forwarded, when you now and then add to your burning Coals, a fat, oily, and fixt Fewel, taken out of the animal or the vegetable Kingdom: For, the Metal must be penetrated by the Phogiston, and receive it both in great Plenty and very intimately: While, at the same Time, the terrestrial and sulphureous Particles which render the Iron crude and brittle, are dissipated: For, the abovementioned Fewel produces Nothing of these Matters, but only constantly supplies a Phlogiston destitute of Acid of Sulphur. Take Care, on the contrary, not to torture your Metal with too violent, too long, and too dry a Fire, nor with an excessive Blast of Wind: Otherwise, you would in vain expect the desired Change.

* Steel Stone.

By

By this Meaus, you will prepare a good Quantity of common Steel, fit for Sale: If any Reader is defirous to have any more particular Apparatus's, fuch as are practifed in feveral Places, he will find many of them, though no Way effentially different among themfelves, in *Swedenborge*'s Treatife of Iron.

CHAPTER SEVEN

On the Metamorphoses of Iron

Pierre Clément Grignon

Extract from the *Mémoires de Physique sur l'Art de Fabriquer le Fer*... (Paris, 1775), pp. 56–90. Translated by P. Boucher and C. S. Smith.

Editor's Note

Literature on the technology of iron and steel increased rapidly in the eighteenth century. A widely known work was the Minerale de Ferro *by Emanuel Swedenborg, which formed Volume III of his* Opera Philosophica et Mineralia (*Dresden and Leipzig, 1734*). *Most of the technical parts of this were translated into French and incorporated in that monumental work on iron, the* Art des Forges et Fourneaux à Fer *by the Marquis de Courtivron and E. T. Bouchu, published in 1761 by the Académie des Sciences as part of the* Descriptions des Arts et Métiers. *This is replete with information on contemporary smelting practice in ironworking districts in France and elsewhere and is especially useful for its description and illustrations of foundry practice. Another valuable source on European metallurgical practice is Gabriel Jars'* Voyages Métallurgiques (*3 vols., Lyons, 1774–81*). *To represent the literature of this period, however, we have selected from the writings of a man who was less eclectic and less learned but perhaps more interesting — Pierre Clément Grignon.*

Grignon (1723–1783) was a thoughtful, practical man, owner of ironworks in Champagne and elsewhere, who deserves to be more widely known than he is. In the preface to his book, he remarked that he knew nothing about the art of iron when he began his career as maître de forge, *but little by little he gained knowledge by continued work, travel, and the interrogation of workmen. The book, he said, was the result of twenty-six years of meditation, observation, and experience, guided by an acquaintance with the principles of chemistry and a taste for natural history. He became a corresponding member of the Académie des Sciences in 1768. Grignon's advice was sought by the French government on a number of occasions, and in 1778 he was sent on a tour of inspection of the whole iron industry in Brittany, Burgundy, Dauphiné, and other districts.*[a] *He also carried out some important archaeological excavations on the site of the Roman villa near Saint-Dizier (Marne), published in 1744 and '75.*

Not being a trained chemist, his approach was too empirical to bring him much acclaim in theoretical circles, but his lively and critical mind led him to a number of acute observations on metallurgical reactions. He regarded the blast furnace containing two thousand to three thousand pounds of iron and the intense heat of the finery as providing better laboratory conditions for the study of the nature of matter and the reactions that modified it than the usual four to six ounces in the crucibles of the chemists. Beginning in 1759 Grignon sent a number of memoirs on iron and on biological curiosities to the Académie des Sciences. These were not printed in the official Mémoires *of the academy, but they were collected and printed together in 1775 as a handsome quarto volume with thirteen engraved plates, under the title* Mémoires de Physique sur l'Art de Fabriquer le Fer.... *It was reprinted in 1807.*

In addition to this book, Grignon's published works on metallurgy include four memoirs on various aspects of the iron industry that he appended to his French translation of Bergman's Analyse du Fer *(Paris, 1783) and his rather copious notes thereon.*

Grignon's memoir on the metamorphoses of iron was submitted to the Académie des Sciences in 1761 together with two other papers on iron, one emphasizing its unity. The Metamorphoses has been selected here because it describes so well the many different forms in which iron appears, not in terms of the best phlogistonic theory of the

---

[a] *Several of the resulting reports and recommendations, which are preserved in the Archives Nationales, have been printed in the excellent study by Pierre Léon,* Les Techniques Métallurgiques Dauphinoises au XVIII siècle *(Paris, 1961).*

day (for the richness of variety was indeed beyond useful theory) but from the viewpoint of an alert, observant, practical man. This, of course, is both strength and weakness. The many changes of iron are puzzling enough to the theorist even today. "Iron is the Proteus of metals." How mysterious the changes in nature resulting simply from the different ways of working iron must have seemed before either the presence of carbon in the products of the blast furnace and finery or their crystalline nature was suspected! And even Grignon with his observations on the different crystalline forms of iron did not think of the most interesting of Protean changes, the internal change from alpha to gamma iron, which is responsible for so much.

Grignon's description of the changes in the blast furnace as the ore descends were not bettered, except for the more precise identification of the nature of the chemical reactions, for over a century. His observations of the conditions under which gray, white, or mottled cast iron are formed are important, even though discussion is in terms of varying amounts of the old-fashioned sulphurous principle and vitriolic salts. We have already met with them in Réaumur and seen their long earlier history. They did not, however, have long to live, for Swedish analytical chemists were soon to identify carbon and silicon. Grignon did see that wrought iron was the purest form of iron, though he believed it to be composed of compacted fibers.

Grignon's observations on crystallization particularly call for attention.[b] Although he was not the first to describe an iron dendrite — that distinction goes to J. H. Zannichelli in 1713 — Grignon is the first to draw one accurately and to speculate on the geometric composition of its branches. He also is the first to give an illustration of the actual shape in three dimensions of the grains in a polycrystalline metal, shapes which do not have crystalline symmetry, although so

---

[b] *For a brief discussion of the development of ideas on the solidification of metals, see C. S. Smith, "The Early History of Casting, Molds and the Science of Solidification," Proceedings of the Second Buhl International Conference on Materials New York: Gordon and Breach Science Publishers, Inc., 1968).*

naturally did he think in geometric terms that he draws them as uniform fourteen-sided polyhedra. Grignon repeatedly urged the consideration of such crystals by scientists, whom he thought had been overly concerned with aqueous systems, and suggested that there was a connection between crystalline form and internal constitution. A statement by Romé de l'Isle in his Essai de Crystallographie (Paris, 1772) *that crystals could only form from aqueous solutions provoked from Grignon a passionate statement that fire provided an equally good solvent and a plea for chemists to study more pyrotechnical reactions.*

Reading Grignon's memoir will provide a vivid picture of iron and steel practices and of speculations on the nature of iron in its various forms in the period immediately before it was seen that simple elemental carbon as an alloying element was responsible for its metamorphoses.

On the Metamorphoses of Iron

or

CHEMICAL AND PHYSICAL REFLECTIONS ON THE DIFFERENT SITUATIONS OF IRON IN THE EARTH, DURING ITS TREATMENT UP TO PERFECTION AND DESTRUCTION; PARTICULARLY ON METALLIC CRYSTALLIZATIONS IN FIRE, ESPECIALLY ON THE CONFIGURATION OF IRON IN ITS MATTE AND IN ITS REGULUS; ON VARIOUS PHENOMENA OF SIDEROTECHNY, AND OTHER PARTS OF METALLURGY.

Non hic vana tenet suspensam fabula mentem.   GEORG FABR.[c]

IRON is a metal of a peculiar species, for in some situations it can be regarded as a semimetal, in others as a metal, and in still others it resembles neither one nor the other.

Cast iron resembles a semimetal[1] in that [57] it is opaque, heavy, sonorous, fusible, and brittle. It differs from the semimetals in that its state is changed by an appropriate operation, and by becoming malleable it becomes a metal, whereas there are no known procedures that confer ductility on bismuth, zinc, the regulus of antimony, or the other semimetals.

Iron is ranked among the metals because of its density, its

---

[1] I do not mean to say that cast iron is a semimetal to the extent that it constitutes a particular species of semimetals. The remainder of the paragraph will show that I give it this name merely by comparison and that its state is not being changed.

[c] "*Here is no idle tale to hold the mind in suspense.*" The German historian-poet Georg Fabricius, 1516–1571.

solidity, and its ductility; but it differs from other metals and semi-metals in that it is not fusible when it has acquired the degree of perfection in metallicity, unless it decomposes.

Iron, properly so-called, in its state of perfection should, when filed, easily acquire a nice polish of a dark gray color; it should be difficult to break; its fracture should be very uneven, dark without luster or facets, and appear to be made up of different bundles of fibers covered with a common envelope that is very smooth, without cracks or fissures; it should heat evenly without much waste and without local scintillations; finally it should [*on heating*] become covered with a golden sweat; inside it should acquire a yellowish-white color, tending toward a pale, non-reddish gold; it should forge without splitting; it should be extendable into either sheets or wires without breaking; and should withstand the hammer while cold, and become springy.

Iron acquires this perfection only by passing through many different kinds of situations in regard to time, quantity and quality of the materials, and the degree of fire used for its perfection. It is the Proteus of metals.

Iron can be considered from five different points of view: the first in its state as an ore; the second in that of cast iron or matte; the third as a regulus; the fourth as a metal; and finally the fifth in its state of decay or destruction.

Iron in its ore is either mineralized or in the form of a deposit that is crystallized, confused, or incorporated with other earthy or rocky substances.

Iron mineralized with sulphur alone (constituting [58] martial pyrites) is, among all the states in which iron is found in the entrails of the earth, the one that gives it the most metallic form; it is as hard as steel and very heavy.

Iron mineralized with arsenic, copper, and sulphur also forms pyrites, the fabric of which differs little from that of the first. Iron combined with other metals in their ores is always in the form of ferruginous guhrs, of encrusted stones, of ferruginous mineral crystallizations, of condensed deposits; and iron that accompanies masses of ductile gold in the mines of Peru is a deposit of condensed ochre, which has no more than a possibility of becoming metal, the same as the ferruginous sands that roll along with flakes of gold in the midst of gold-bearing rivers.

Iron ores deposited by the dissolution [*weathering*] of pyrites are more or less pure and more or less rich depending upon the circumstances that accompanied them. It is the same for rocky and earthy substances and for all sideroliths. I shall not linger on this topic; I have already spoken about it, and many more volumes would be necessary to make it clear. I shall pass to cast iron, which is the second state of this metal.

To obtain iron in the first form in which it figures in commerce and the arts, the ore, adequately cleaned, is placed upon charcoal[2] that is thrown into the smelting furnace at regular intervals of about 80 minutes. During the first period, or the time that it takes for one charge to be consumed, the ore loses its humidity and descends about 36 inches. During the second, the ore receives what may be regarded as a roasting heat which removes some of the most volatile foreign parts; it descends 30 inches. During the third period, the ore becomes red, as does the calcareous and fusible matter [59] that accompanies it; it descends 22 inches. During the fourth, the ore sweats out some matter that is the most fusible part of the foreign bodies that it contains and which makes its pieces adhere to each other; it descends 18 inches. During the fifth period, the ore softens, takes on phlogiston and, without changing its external form, it receives the first degree of metallization; and it descends 15 inches. During the sixth, it enters into a thick, mushy fusion which balls up and amalgamates with the flux which has become lime; and it descends 12 inches into the great hearth of the furnace (the area in which the two cones — that of the walls and

---

[2] I will give the dimensions of the interior of a furnace for smelting iron in the Memoir on the art of economically washing and smelting iron ores. [*Memoires*, pp. 91–183.]

that of the hearth — are united at their bases) where the materials combine, and where there dwells the greatest ambient heat. During the seventh period, the union of the materials becomes more exact; fusion increases, the sulphurs develop and part of them reacts upon the ore and melts it, and part is absorbed by the calcareous materials, as well as to some extent by the absorbent earths; the ore descends 8 inches. During the eighth period, purification begins to take place; fusion reaches the point of forming globules which escape in sparkling drops; and the charge descends 15 inches to find itself above the tuyère. Finally during the ninth and last period, the ore descends imperceptibly from the hopper, at the rate at which the charcoal is being consumed, opposite the tuyère where it is intimately penetrated by fire. This causes the phlogiston that it has received to react upon its molecules and divides them to the point of rendering them fluid. The action of the fire at the center of the hearth is so vehement that vitrification of the materials is instantaneous; the cast iron therefore enters its degree of fusion, penetrates the vitrified mass or the slag that is beneath it, deposits whatever [*nonmetallic matter*] is attached to it, and lies in a bath at the bottom of the hearth. It is in continual internal movement there, from which purification results. Finally after the twelve hours required for the reduction of the nine charges, the cast iron is either tapped or [60] withdrawn with ladles, depending upon the different uses for which it is intended.[3]

Cast iron, or rather iron matte which is its proper name,[4] leaves the furnace in different stages of purity, consistency, and color and with different properties. We ordinarily distinguish two

---

[3] The duration of a charge, the space it occupies in a furnace, and the changes it undergoes are somewhat dependent on the construction of the furnace and the manner of operating it.

[4] Some authors give to cast iron [*fonte*] the designation of melted iron [*fer fondu*]. This expression is all the more improper as iron, properly so-called, does not melt; if it has been melted, it is neither iron nor is it cast; and cast iron has not yet acquired the state of iron with its properties. [*This might have been clearer had we translated* fonte *as pig iron, but when molten it is no more porcine than it is cast!*]

different kinds, one white and one gray, each being subdivided into various nuances of shade and quality.

In general, cast iron is nothing but the smelted ore that has retained part of the sulphurs and other substances contained in it and which was charged, during fusion, with a superabundance of the volatile sulphurous principle of the coals; this reduces it to a more or less pyritic state. Cast iron is as different from iron as antimony is from its regulus.[d]

White cast iron, which is said by one author to be the best[5] is, on the contrary, the worst because it is the one most highly charged with heterogeneous matter. White cast iron leaves the furnace thus for several reasons. First; when the furnace is overcharged with ore in relation to the heat that is available — either because the heat is insufficient on account of faulty construction, weak action of the bellows, or poor quality of charcoal, because it was burned too rapidly, or because it was rotted in the store;[6] these accidents [61] all oppose the correct departure of foreign matter, which remains intimately united to the cast iron.

The second cause: when a founder is not careful to work his operations so that the charges descend gently, it happens that the materials form a vault ["*scaffolding*"] over the tuyère, and when the parts that formed the arch become loose, the charge collapses and falls confusedly. The smelting cannot be correct because the heat is too divided; the materials are thrown into the bath before they

---

[5] Monsieur de Réaumur took the reguline part of iron to be cast iron. Metallurgical work on a small scale often takes the shadow for the truth. [*It is, however, true that white iron contains smaller amounts of foreign elements than does gray iron.*]

[6] Rotting is not taken in the strict physical sense, to express the destruction of molecules of charcoal, but to express the accident that happens to it because of the abundance of moisture that can penetrate it; and which, rendering it very massive, slows the development of the burning parts. I have some charcoal that is enveloped in a peculiar fungus. This is not the time or place to analyze all the causes that astonish the physicists, charcoal reputedly being indestructible by any agent other than fire.

[d] *The sulphurous principle is the reducing principle, a synonym for phlogiston, though Grignon here is confusing it with the "real" element. "Antimony" is the sulphide (stibnite); metallic antimony is the regulus.*

have been sufficiently prepared and carry with them the heterogeneous bodies that did not have time to separate. The same accident occurs when the furnace walls bulge out too much; materials collect in the corners of square furnaces[7] which accumulate to the point of forming masses whose weight suddenly precipitates them into the bath. A similar inconvenience occurs when an old hearth has been too much enlarged: it can no longer support the balance of the column of materials, and everything is then mixed up together. Finally, all the causes that can diminish the heat of the furnace interrupt its operations and impair the resulting product. It is a stomach, well or poorly constituted, that reacts differently upon the different materials that it receives. When the food is not prepared by selection, maturity, cooking, mastication, or is given too precipitatedly or in too great abundance, or is of a harmful quality; or because the muscles have lost their resilience on account of decay, the digestion can therefore only be imperfect; the chyle that results is bad. A foundryman, or rather the master of an ironworks, must carefully regulate the diet of a furnace in order to prevent sickness and to guard against the numerous accidents that take even the most vigilant people by surprise. My present topic does not deal with the prognosis and diagnosis of the state of a furnace: that [62] will make up part of another Memoir. I shall just mention that all the accidents cited above give a more or less white cast iron, of which there are three kinds.

The first kind comes from violent disturbances in the furnace; it emerges in a mushy state, troubled internally by the effort of the

---

[7] In order to prevent this accident, I make the walls of my furnace very steep, and the interior of the whole hearth is constructed on elliptical lines, contrary to the opinion of Orchal. *Grignon's elimination of the corners in the furnace hearth was an important improvement which was soon adopted in all iron-making districts. The Orchal referred to is J. C. Orschall, whose* Ars fusiora fundamentalis et experimentalis, *first published in 1687, and other metallurgical writings had appeared in French translation in 1760.*]

foreign materials to release themselves, forming bubbles from which leap jets of flames; it is heavy, brittle, dark on the outside, often reddish, white inside, without luster or order, becoming red at the fracture if broken while still hot; it has a hard, harsh sound. This particular cast iron, which should be called iron matte, is useless for any purpose in this state: at the finery it requires a good deal of work even to be made into very bad iron.

The second kind of white cast iron is that which is so due to some slight accidents or because the proportions of ore and charcoal are maintained in such a manner as not to allow more exact purification. This cast iron, by reason of the great quantity of foreign metallic and sulphurous parts that it contains, attacks and corrodes the crucible that receives it; it emerges impetuously from the furnace, in a fiery state; it boils, shooting off clusters of flames that are very pleasant to see at night; it congeals very promptly with an uneven surface; it is covered with a hard, black, brittle crust that comes off in scales (Pl. V, Figs. 1, 2, 4. [*sic. None of the plates shows the scales.*]) The interior of this cast iron is very white, arranged more or less regularly radially, like crystallized martial pyrites or antimony, or rather like all metallic substances that are intimately united with much sulphur. This cast iron breaks with a loud crash on cooling if its volume does not have a thickness proportionate to its extent, as in the case of plaques, chimney-backs, pots, and other things of this kind; and when it is in masses of considerable size, such as hammers, anvils, and so forth, it breaks in use in proportion to the strain it undergoes. This cast iron has a clear silvery sound and could well be [63] used for the making of bells. It is heavy, hard, and brittle. The file does not begin to cut into it; even the hardest files with large coarse teeth scarcely have access for splitting it; the abundance of foreign and sulphurous materials makes it yielding to fire; it is very fusible; easy enough to work in the finery, though it gives a [*wrought*] iron that is fugitive, breakable, hard, hot-short, and compact. In general the use of white cast iron should be prohibited for working in forges. White cast iron is capable of being purified, but this doubles the labor and expense. It is white and of a compact texture on account of the insertion of foreign bodies between its molecules; it is this contiguity of parts that renders it sonorous. Its radii [*radial crystals in spherical castings*] are tetragonal prisms, very slender and made up of rhombs placed upon each other. (See Pl. V, Fig. 3.)

The third kind of white cast iron is that which has received a higher degree of purification than the former; it is a little more perfect, though it still contains heterogeneous, sulphurous materials, and partakes of the quality of gray cast irons: this can be ascertained from particles of the latter, that are more or less abundantly scattered throughout its mass, and which form starlike gray spots, somewhat similar to the spots on dogfish and trout; from whence its name, mottled [*truitée*] or mixed cast iron. This kind belongs more to white cast iron than to gray because the latter is always less dominant.

Mottled cast iron leaves the furnace in a more liquid state than the former and more gently; however, it shoots off brilliant sparks[8] which disclose its quality and its imperfection. If the work of the furnace has not been too hasty, this cast iron can serve to make large objects [64] such as balance weights and the like. It is very proper for anvils, for forges, and for all sorts of things where volume contributes to strength. Refiners prefer it to all others because of the ease of working. It gives better iron than the preceding ones, even in proportion.

There is a fourth kind that can be placed among the ranks of white cast iron. Although it is so merely by accident, it belongs here because its causes will throw much light upon the nature of white cast iron in general.

When cast iron that is by nature gray is received in a cold,

[8] These sparks are globules of cast iron, hurled by the rarefaction of foreign bodies and the air that is released by the great heat. These globules sparkle because, being in free air, their phlogiston leaves them in the same manner as when one strikes a tinder-box.

humid, compact body, it congeals precipitately and becomes white, hard, and brittle, so that if a piece is molded in such a manner as to make it unequal in its thickness, even though it is cast from the same drop of gray cast iron, the thinnest part is white, that which is a little thicker is mottled, and that which has the greatest volume is gray; a singular phenomenon of which I shall try to develop the causes.[9]

We cannot doubt that cast iron, in whatever condition it is found, even the most perfect, contains many sulphurous superabundant parts; that the more it retains, the less perfect it is; and the easier it is for them to escape, the purer it is. This separation is always effected in proportion to the degree of heat applied.

A cast-iron stove heated solely by the fire that it contains spreads an unpleasant sulphurous odor within the drying-room that it heats. The violence of its effect depends upon how ardent is the fire, how new the stove, and how well the place is closed up. Inasmuch as an earthenware stove does not produce the [65] same effect or to the same degree, this occurrence must be attributed to the escape of the sulphurous principles contained in the cast iron of which the stove is made.

This exhalation of the sulphurous principle is so considerable in cast iron in fusion that if several loam molds of different objects are buried in the same pit, all without communication and even separated by a partition of ten to twelve inches of hard packed sand, and if cast iron has been introduced into one of these molds, the neighboring molds are so penetrated with the sulphurous matter that emanates from the cast iron in fusion and the air is so rarefied that if a light is brought close to a vent of one of the empty molds, the air bursts into flames and flames come out continuously.

---

[9] Monsieur de Réaumur did not understand this phenomenon, and because he attributed superiority to white cast iron, he regarded this essential defect as a perfection in order to make it compatible with his reasoning. [*In the second edition of his work*, Nouvel art d'adoucir le fer fondu (*Paris, 1762*), *page 80, Réaumur did correctly explain chilling:* "*White iron is nothing but quenched cast iron.*" *This was written before 1726 but not published until 1762, after Réaumur's death.*]

This precaution in setting fire to the molds is absolutely necessary; for if cast iron is introduced without this precaution, the explosion is so considerable that it fractures the cope and scatters the cast iron dangerously.

If large plaques are to be cast on sand, air channels called vents are made under the mold; when the iron is cast, fire is produced and there comes out a burning air, often sparkling. Without these vents, the air and the volatile sulphurous principle, finding obstacles to their escape in the mass and compression of the sand, are constrained to force their way through the cast iron by making cavities and blowholes, or the plaque breaks with a noise on cooling.

Finally, all masses of cast iron that are cast in molds whose porous matter allows the escape of this superabundant sulphur are surrounded by a burning, bluish atmosphere that continues long after the solidification of the cast iron.

It is not the same when cast iron is introduced into molds made of a substance (too compact, too cold, or too humid) which impedes the escape of the sulphurous principle contained in the iron. Examples are all [66] molds of cast iron, such as bullet molds, poorly baked loam molds, and molds of sand that is too wet. Gray cast iron introduced into these comes out white because in poorly baked or in damp molds the outer parts of the cast piece are suddenly congealed. When the piece is thin, as in kettles, cooking pots, pans, and other objects of this type, the interior is also soon congealed. The sulphurous principle remains united to the cast iron which, instead of being soft and gray, is hard, tin-white, and fragile enough to break by itself; it becomes honeycombed inside; it is of the greatest importance to take the most carefully considered precautions to prevent this occurrence, especially for pieces of artillery. (See Pl. V, Fig. 5.) A bomb bursts in the mortar because the explosion of the charge produces some openings that carry the fire to the interior of the bomb; or it does not describe the parabola necessary for its trajectory because it does not have the proper weight in relation to its diameter on account of the holes scattered throughout the mass. Cast-iron cannons burst for the same reason: these should always be cast without a core and then bored because the core is liable to be displaced, consequently causing errors in the thickness; or if the core has not been well baked or if it contains bodies upon which heat or cast iron react, there results a rarefaction that fills the core of the piece with holes. Cannon balls that are ordinarily cast in chill molds of cast iron are liable to be hollow because the cold and the close texture of the chill mold do not permit the escape of the sulphurous principle and of fixed air; they work inside at the center of the mass, causing there a boiling that scatters and rarefies the cast iron, thus producing holes. Such a cannon ball fills the bore, but it does not have the proper weight in relation to its diameter. (See Pl. V, Fig. 4.)

I do not pretend to attribute all these disorders solely to the escaping of the sulphurous principle. Air released [67] by the materials that receive the cast iron and rarefied by the great heat also plays a great part; it is fixed air combined with humidity and with a very great abundance of sulphurous matter.

This would be the place to talk of cracks and fractures in pieces of cast iron, an event that occurs long after they have been cast and is often caused by a little ray of sunshine after a thunderstorm; of ringing sounds in chimney-backs, hearths, or stoves that are heated hot in very cold weather; of the fracture at the end of pigs of white cast iron in the finery hearth, and of the means of preventing the accident that has much to do with the system of electricity; but these details, which will have to be examined more thoroughly in other Memoirs, would plunge me into extensive reflections far removed from my present subject.

I shall finish the story on white cast irons by saying that they are so only because they contain many foreign heterogeneous, superabundant, and particularly sulphurous materials, in greater or smaller quantity depending upon whether they received more or less heat in the furnace; that they were congealed more rapidly;

that gray cast irons that are whitened by accident have acquired this defect by the sudden cooling that intercepted the escape of the sulphurous material and gave them a quench that disturbed the order of the symmetrical arrangement of their parts, and gave to them something in common with steel, in addition to hardness and brittleness; that in general white cast irons crystallize in concentric layers, like all metallic substances combined with sulphur; that we can take away or give them this sulphur, and consequently that white cast irons are nothing but a mineralized iron of the second kind, taking martial pyrites as mineralized iron of the first kind.

Gray cast iron is that which is obtained by a proper proportion of ore, fluxes, correctives, and heat; from whence follows the departure of the [68] heterogeneous materials that are vitrified and an exact fusion of the metallic parts. It is this kind of cast iron that produces the best [*wrought*] iron, so that it is possible to extract good iron from bad ores by seeing to it that they are reduced to gray cast iron.

In general there are two kinds of gray cast iron, one ash-gray, the other much darker, tending more or less to black. The former is that which is in its state of perfection, considering it as cast iron useful to the arts. It comes out of the hearth as fluid as water seeking its own level. It is not to this cast iron that one should apply Geber's definition, *non fusibile fusione recta*; for holes made accidentally in the furnace outlet by the tapping bar give jets of fire that rise nearly to the level of the bath and produce a marvelous effect at night. This melt is quiet, of a beautiful golden yellow color, glistening in the sun; it flows back and forth when poured into a horizontal mold; exhales some yellowish-white vapors; takes all sorts of impressions, even of most delicate chiselings in the hands of a skillful molder; has a very considerable shrinkage,[10] and is covered on its exterior surface by a very light

---

[10] The authors who have asserted that cast iron has no shrinkage were duped by the imperfect fusion in their experimental furnaces.

skin of slag. Its external color is a slaty gray, bright when it is raw and silvery when it is polished. It rusts with great difficulty on the exterior but very promptly on the interior. When broken, it is bright ash-gray inside when it is at the peak of its perfection, when it has not received a degree of hardening by sudden cooling, and when the eruption of superabundant sulphurs, rarefied by heat, has been facilitated. It can be filed, is difficult to break, has even a little elasticity; the chisel can cut into it, and the hammer makes impressions by compressing its molecules. The [69] shape and arrangement of its inner parts depend on the circumstances that hastened, slowed, or prolonged its cooling. When some cause has disturbed the order, the arrangement is confused, with a steel-like grain more or less large and rounded. Very slow cooling confers a symmetrical arrangement on its molecules. This is what I am going to examine.

There is nothing in nature that is not characterized by an essential individual form over which chance has no dominion. Every being has a determinate characteristic shape which, in concert with the equally invariable quality of its substance, determines its properties. It is a triad in which I conceive that the shape proceeds from the essence of the substance and the properties from these two. They came into existence at the same moment, for matter cannot exist without form, and it acquires properties at the instant that it takes form.

All bodies that are capable of receiving the consistency of a fluid by Nature or Art respectively take an essential symmetrical form on condensing. Each molecule similar to its neighbor approaches and unites with it, and successively, as the fluid that divided them is dissipated, they shed their obstacles and tighten the bonds of an invariable affinity. They attach themselves to each other in numbers always related to the number of faces and to the span of the angles of each molecule. That is why, when I see a body that is naturally cubic, I conceive that each molecule of this body is a cube; a rhombohedral body is composed of rhombs, and so for the others.

(Pl. V, Figs. 6, 7, 8, and 9.) Spar and quartz, which play such large roles in the mineral kingdom, give proof of what I have just advanced. Are not all precious stones, which are for the most part metallic crystals [*i.e., crystals containing metal in their composition*], ordinarily found in a concrete, regular form?

Saline, acid, and alkaline substances dissolved [70] in water, bitumens dissolved in oils, metals dissolved even in acid spirits — all these condense into regular essential, and invariable shapes. These configurations have all been described, but nothing has yet been said of metallic crystallizations occurring in fire.

Would metals, which are the noble parts of the entrails of the earth, be deprived of a property common to all existing things? Would the shape of their parts submit to the whim of chance and of the workman who manipulates them? No, without a doubt they have a generic, differential form, in accord with the principle that I stated above, and are subject to all the laws that follow therefrom.

All metals, when dissolved by acids, give crystals of a determinate compound shape which proceeds from that of both the acid and the metal: they can be called hybrid crystals, to distinguish them from natural crystals. There are some authors who have described several of these hybrid crystals very well; others, lacking terminology and carefulness, have said that they were in [*the form of*] needles.

We have seen, as an inevitable consequence of the essence of things, that white cast irons crystallize in slender prisms placed in converging radii. This state of affairs can be varied by cooling because the point at which cast iron first cools is that at which the first condensed molecule attaches itself; and so forth. (See Pl. V, Fig. 2.)

A cannon ball begins to cool at its circumference. The molecules that are nearest by attach themselves, and as the heat withdraws to the center, the molecules accumulate one upon another, following the progress of the cooling, finally up to the center, which is the point toward which they all tend. If for some reason one side is cooled rather more than the other, all the radii will be directed at angles in relation to the position of their bases, so that those on the side that first becomes cold will be the longest, and the opposite ones the shortest (Fig. 1, AB) I have assumed that the form of the [71] crystals of white cast iron was determined in part by the sulphur that it contains. Let us remember the internal shape of pyrite, of antimony [*sulphide*], and cinnabar. Therefore as cast iron frees itself of its superabundant sulphurs, it must take a different arrangement during its crystallization: the fact confirms the reasoning.

Gray cast iron, in its degree of perfection, gives a very regular crystallization, each crystal being distinct and isolated. To obtain it, however, the cast iron must cool very slowly for several days, the shrinkage must be considerable, and nothing must disturb the order: therefore each crystal is a kind of pyramid whose base is a rhomb, having other pyramids applied at right angles and continually along each face. The base of these pyramids, at the point where they meet, is equal to the diameter of the principal pyramid to which they are attached. Since the diameters diminish successively, the pyramids at the bottom are larger and longer, those at the top shorter and more slender, an appropriate proportion being maintained between the diameter of the base and the length of the column. The four pyramids that are opposed, in the form of a cross, are in every way equal to each other. Successively they come to meet, by regular decreasing of spacing, length, and thickness, following an imaginary oblique straight line at the top of the central pyramid, whose very acute apex is an aiming point from which can be seen all the points of the lower pyramids; these constitute each crystal of cast iron, regularly grouped. The little grottoes, where masses of these crystals were formed, offer to the eye aided with a magnifying glass the spectacle of a little metallic forest made up of trees with opposed fourfold branches. Each pyramid is made up of a succession of rhombs whose sides are inclined. This provides a larger surface that is applied upon the smaller one of the rhomb that supports it; and so on successively. These crystals are larger and more depressed than those of white cast iron. They are isolated because they proceed from a [72] more homogeneous cast iron and because its perfect fusion favored the exact arrangement of its molecules. (Pl. VI, Figs. 11, 12, 13, 14.)

Gray cast iron is much less sonorous than white cast iron because its parts are far more continuous, and they are more supple. This is why charlatan salesmen who have worthless pieces of white cast iron make them ring to obtain a sale. It is a snare that traps most housewives and ignorant persons: *Ferro fusurae, nimium ne crede sonoro.*[e]

Little bells are more sonorous in relation to their mass than large ones, if their ingredients are in similar proportions, because the lesser volume of the small ones carries less heat during casting. Because the molecules of the matte condense more promptly and more confusedly, they are less continuous and consequently more susceptible to vibration. Instead, the monstrous bells from Peking and so many others in Europe surprise us by not producing the sound that we have a right to expect from the enormous volume of their mass. I find that the cause lies in the prolonged cooling of their matte. During the long span of time required to congeal forty to eighty thousand [*pounds*] of molten matte, the molecules, particularly those at the interior of the metal, unite intimately and continuously, acquiring their natural configuration. Each crystal of metal is found to be more or less separated from its neighbor because there are necessarily an infinite number of small voids, as the [*solid*] matte is of a tighter, more compact texture than when it was fluid. The vibration is therefore less sudden and less general, inasmuch as each molecule performing the office of a spring hammer strikes its neighbor when it is set in motion; if it is off balance, not only does its vibration cease because it has no reaction but also it transmits no movement because it met nothing in its

[e] "*If iron is sonorous, do not trust it for casting.*" Author unknown.

fall: [73] it is a semitone. The intelligence of the workman should suggest means of finding a remedy for this kind of defect, either in the shape of the bell or in the proportions of the mixture of its ingredients. But most of the founders who run around the countryside appeared to me to have so little instruction and so little precision that I am not surprised they often meet with such little success in their operations. I return to my topic.

When ore has been used too sparingly in a furnace, or the degree of heat increased by a more vehement blast or a more generous use of charcoal, or, finally, if the cast iron has remained too long in a hot bath, it is then very dark gray, often blackish. This is the second kind of gray cast iron. This cast iron leaves the furnace sedately because its too great concentration is an obstacle to fluidity. It is dull, covered with wrinkles that are formed by the folds of a film composed of its own substance, which promptly loses its fluidity at the surface; the movement of air, excited by the heat, carries along particles that float and sparkle in the air. These small bodies, merely a very attenuated form of cast iron, are called kish,[f] and the cast iron that produces it is called kishy iron. Its texture is sparse, which renders it less heavy. It is soft to the file. Chisels can cut into it; but it breaks up into grains more easily than gray cast iron. It can endure a violent force before breaking. It is very hard in the fire, requiring laborious attention at the finery, but it gives a fibrous and compact iron. The configuration of its crystals is the same as that of gray cast iron; but they are shorter and grouped irregularly on account of the confusion that arises from its too rapid solidification.

This cast iron suffers a very great waste in the yield of a furnace. Objects of small volume that are cast from it are rarely successful because the kish that it forms prevents it from taking the impression

---

[f] *French* limailles, *literally "filings." Though Grignon grasped the origin of this material, he misunderstood its nature. Kish is graphite crystallized in the form of flakes during the cooling of carbon-saturated iron. The flakes are not wet by the molten iron and so are released into the air.*

of the molds [74] and from uniting perfectly, so that the objects are scaly, wrinkled, and often contain holes. Pipes for conveying and raising water should not be cast from it, for the water would seep through. This cast iron is very proper for casting thick pieces that need great resistance, such as shaft collars, trunnions, bearing blocks, and cranks for large machines, as well as for all objects that are to be polished on a lathe, or on which one would like to finish off round relief by chiseling after the cast iron has been appropriately annealed.

Kish is an undesirable complication when one wants to cast objects of the kind that it ordinarily ruins. To rectify it, an intelligent founder who is alerted by the color and consistency of the slag, even by the kish which adheres to his tools during the work, an hour before casting cautiously introduces pieces of white cast iron in such quantity as is needed, or a bit of lead. The latter method succeeds very well when [*the lead is*] put in the ladles with which the iron is withdrawn. In either case, the kish is revivified, and the objects succeed because the methods employed furnish sulphur and phlogiston. Kish is therefore cast iron that has lost almost all its superabundant sulphurs and a bit of phlogiston and is no longer fusible unless these be restored. Everything that contains some will make it return to its natural state. The more the [*furnace making the*] cast iron is blown, the more kish it gives. When after a shut down[11] there remain a few coals and a bit of cast iron and these are blown cold[12] to accelerate the cooling of the furnace, the cast iron that is found to be lodged in some confused masses of slag and charcoal is converted into kish, a mass of small, very slender platelets of a bright black color, more or less attenuated. [75] The thinnest and smallest among them are silken, greasy to the touch, and they blacken the fingers when they are crushed. They remain suspended in the air for some time, pleasantly reflecting the light of a beam shining across a dark room. It resembles ferruginous

---

[11] After the work in the furnace has been finished.
[12] This is, working the bellows without material.

mica, which following the common terminology one might call cat's [*eye*] iron. It also closely resembles a bright black iron ore from the Isle of Elba that is treated in Corsica. These small platelets are so many attenuated particles of the regulus of iron, which is the third point of view from which I consider iron.

When cast iron remains in a bath for a long time beneath a layer of material that is capable of preventing the loss of its essential principles while permitting the escape of the heterogeneous materials that it contains (or even absorbing them), the cast iron then condenses into a compact, hard, brilliant, silvery material, crystallized in the form of hexahedral rhombs, cubes, [and] parallelepipeds composed of a fabric of layers laid upon each other. These can, with effort, be broken into rhomboids, in the same way as Iceland spar, each sheet being made up of intimately united rhomboidal molecules, whose separation by fire forms the kish that I have just been talking about. This kind of crystallization is very difficult to obtain regular. It is ordinarily confused. It occupies the mean between the state of cast iron and that of wrought iron: it is the regulus itself which is very slightly malleable and which no longer melts completely. For if it is heated, it passes through different states, or it mineralizes, becomes clinker, or it converts to ferruginous amianthus, as I demonstrated. (See Pl. VII, Fig. 19, RO.)

I am using the term regulus in a general way. I adopt the opinion of Becher who defines it thus: *Chaos seu medium inter mineram & metallum; quod nec corpus nec spiritus est, sed quoddam mirabile quod se ad metalla habet, ut mater.*[g] I have not found a name more suitable to this substance in which the ferruginous part is not [76] associated with pure sulphur as in pyrite, is not decomposed as in ores by erosion and deposition, is not mineralized to the point of fusibility as in the various kinds of cast iron. But by the purification caused by prolonged fusion, it has been reduced to the point of fixity of not being susceptible to fusion — *nec spiritus* — while

[g] "A chaos [i.e., disordered state] or halfway between a mineral and a metal; it is neither body nor spirit, but holds itself as a mother to metals."

still containing heterogeneous parts interposed between its molecules and preventing ductility — *nec corpus.*

When this regulus is cooled rapidly, its crystals are so small that they acquire a steel-like grain, and it is then analogous to the alleged native iron of Germany, of which I have seen many pieces both fractured and not broken.

Regulus of iron, passed through the refining operation, gives a soft, fibrous, and compact wrought iron; while wrought iron made with ordinary cast iron, produced by the same ore, gives iron that is brittle, rough, hard to the file. In accordance with this principle, if we want to purify cast iron in order to obtain better iron, we pass it through a refinery fire;[h] after it has remained in the bath for a sufficient length of time, the purified material is released through the taphole and is received in a space limited by a bank of sand so that it forms a parallelogram. Water is thrown on the iron to detach the slag that covers it. Water hardens the slag and rarefies the iron, so making a wedge between the two substances and separating them. This refined iron is cut up by marks drawn transversally with a piece of wood. It is somewhat like working cakes of rosette copper.

The cast iron that came from this melt approaches the state of regulus more or less exactly in its degree of purity; it is not exactly fusible anymore, and it can hardly even be introduced into molds, especially if it has been purified too much. That is the reason why the little pocket-edition founders who remelt cast iron to make pots with it use the term wild [*mad*] for cast iron that has already been remelted several times and is no longer capable of fusion. The more they give it a violent and continuous fire to [77] get it into fusion, the lumpier it gets, because this last fire, which they try in vain to give it, is a refining fire that metallizes it.

[h] Feu de macération. *This is the preliminary fining operation under an air blast, which had the effect of removing most of the silicon but little carbon and so simplifying the work in the next stage. Chemically it is equivalent to the refinery process that was later used for pretreating iron for puddling, and to the first stage of the Bessemer process.*

In work on a large scale, the boiling caused by the dampness of the sand upon which this refined cast iron was cast, and of the water with which it was sprinkled, renders this reguline cast iron more accessible to the fire because of the numerous holes with which it is pierced; this makes it resemble in every way the alleged native iron of Senegal, which has absolutely the characteristics of refined cast iron that it received from an existing or extinct volcano.[1] The testimonials of men born several thousand years after this accident cannot prevail against the evidence of Nature. The principles of iron are so coarse that a violent fire is needed to penetrate them. Their metallic aggregation cannot be effected by the repeated acts of imbibition of earthy juices, whether aqueous, saline, or sulphurous, nor by the slow and successive penetration of mercurial vapors, none of which has access to them. Furthermore, the juices that condense almost all other subterranean bodies would be more capable of destroying the metallic state of iron than of producing it, since the aqueous ones, by carrying along the martial particles, make up the ochres and sedimentary ores; the salts [*give rise to*] vitriols; and the sulphurs to pyrites. Mercurial vapors, which vivify nearly all other metals in their matrices (whenever the circumstances allow), are not of the essence of iron, are not even capable of union with the substance of iron: therefore all these general agents cannot contribute to the formation of virgin iron. The consummation of this operation, prepared by Nature, is reserved to the most violent fire found in volcanoes or in furnaces that have been built to simulate them.

I am convinced of the possibility of finding fossil iron like fossil wood. But should one conclude from the existence of the latter that vegetable growth to form subterranean forests could have taken place in an inverse or horizontal situation, and without the help of free air? That Nature would have strayed from her

---

[1] *Grignon is probably referring to an iron meteorite, for many of these have wormholelike cavities on their surfaces strongly reminiscent of the large blowholes in refinery iron or wash metal, although their origin is quite different.*

immutable laws to produce a unique phenomenon? [78] The absurdity of such reasoning would be as monstrous as proving that sharks and whales are androgeneric because whole men are found in the caverns of their stomachs.

In the fabrication of iron, cast iron is generally not purified by refining [*maceration*] before it is submitted to the finery. This operation is reserved for the hammer forges that make either steel or iron intended for ornamental objects. In nearly all other large-scale work on iron, cast iron is submitted immediately to the finery to make iron in a single operation, so that the fire, being applied to it less repeatedly and for shorter times, has less of a hold on the heterogeneous materials contained in the cast iron, and a more or less abundant portion of these passes into the masses of iron that leave the fineries: these masses contain a greater proportion of foreign materials the larger they are, because the surface presented by them to the action of the fire was less.

Iron, on leaving the finery, is therefore nothing but a bloom of ferruginous matter, more or less pure, whose parts are separated from one another by a vitrified matter, a portion of which is pushed out by the pressure of the hammer, which brings the molecules of iron closer together: these assume different shapes and weld together more or less precisely in accordance with their degree of homogeneity. Those that still contain a superabundant foreign body clump together and crystallize more or less regularly. Those that are entirely deprived of all parts foreign to the essence of the metal are attenuated to the point of forming flexible fibers: each molecule of iron draws closer to its analogous neighbor and fortifies itself.

If we break a bar of iron, it is brilliant either along the entire extent of its fracture, or only in part, or finally not at all, but on the contrary it is of a dark color. When it is brilliant over the entire extent, [79] it is so because of the reflection of an infinite number of facets that occupy all the circumscribed space. These facets are either very small and multitudinous, or are more extended and less numerous, or finally they are very large, few enough to be counted and to show their dimensions. These three situations of the molecules of iron are three degrees of imperfection that remove it proportionately from the state of perfect metallization and prove that it is more or less reguline. The larger the facets, the more reguline and imperfect the iron; the smaller they are, the farther removed they become from the state of a regulus, the more adherence they have and the more they tend to perfection.

A glance will prove that coarse iron is nothing but a regulus, since these facets are only broken crystals of regulus.

When the end of a roughly forged bloom, which is a crude form of iron, is submitted to the chafery to a degree of heat sufficiently high to cause the slag that it contains internally to enter into fusion, enabling the molecules of iron that were united to it to come closer together, they take a regular form in this slag, which was its solvent and which serves as the vehicle for its crystallization. If the heat is forced and it bursts under the hammer, there escape groups of those crystals that the smiths call grains [*grumillons, literally little clots or lumps*]. (Pl. VII, Fig. 20.) If, on the contrary, the heat is adroitly gathered together, the slag transpires through the pores, the crystals become irregular from the pressure they receive when in their soft state, and they unite on all sides to form a bar which, when broken, presents brilliant faces, angles, and cavities that are nothing but the surfaces of these crystals seen in a different aspect.

In masses of chafery slag I found considerable groups of these crystals formed by parts of iron that had escaped with it through the taphole and which had crystallized in the slag (which had fragmented) as in their natural solvent. Slag [80] is to iron crystals what water is to crystals of salts: it is the solvent from which they precipitate under their essential form, a portion of which they retain and which links their symmetrical arrangement. Just as the mother liquors that remain after crystallizations of salts contain a portion of acids or alkalis and their bases changed by multiple solutions and by the attenuation that their molecules have undergone by repeated application of heat, so does slag contain a residue of the parts of iron that can no longer make a metallic union, even though they are attractable to the magnet.

These iron crystals are rarely very regular because the fire which gives birth to them welds them together and mutilates their angles by the action that it has on their substance. The most regular appeared to me to be polygons, hexahedra, formed from several rhomboids united by their large faces. (See Pl. VII, Fig. 17.)

On breaking a bar of iron that is more purified than that which I have just described (that is, one that has acquired a degree of perfection above reguline iron), the fracture will be strewn in some areas with very small facets, called grains, and with sinewy tufts which prove that the iron has very nearly attained the point of perfection. It is held back by the lesser or greater abundance of these small grains that are still impregnated with foreign materials — the departure of these makes a noble iron that is bristling all over its fracture with inequalities of a dark gray color. These inequalities are made up of a fabric of sinewy fibers which, applied one upon another, constitute muscles in the iron that give it strength and elasticity. These fibers are more or less refined. Occasionally we notice small clots that interrupt the parallelism of the situation of the iron fibers; these are so many little imperfect parts.

These fibers (which appeared to me to be capillary, that is, tubular, when in their state of perfection) are the [81] essential homogeneous parts of iron. Any iron in a natural situation that is not made up entirely of fibers is not [*pure*] iron, but is reguline iron, more or less remote from its state of perfection. I say in a natural situation, for the work by which iron is made into steel merely tends to interrupt the continuity of these fibers and to give them rigidity, which makes me say (until I have acquired more light on the nature of steel) that the latter is iron arranged in an unnatural situation.

Iron in general, and more so in its state of perfection, has the properties of an organized body, so to speak, since it contains within itself a subtle, fluid material that is capable of circulation when it has movement communicated to it, either naturally by a position directed toward the [*earth's magnetic*] pole, or by rubbing together two pieces of iron in the direction of their fibers, or by rubbing with a lodestone. It also has the property of strongly attracting, containing, and transmitting the electric fluid, of annihilating the effects of lightning, so that lightning has never been known to fall on an ironworks.

After having attentively observed all the numerous situations of iron since its existence in the womb of the earth and the different shapes that it assumes as a result of the effect of fire which confers on it its different degrees of purity, I conclude that there is no metal that varies more than iron during the operations of nature and those of the art by which it is carried to its perfection; that the parts of iron, following the natural order of things, have a regular, symmetrical, essential, and characteristic shape; that all metals must similarly acquire a distinctive and relative shape; and that in order to be convinced one need only to make them enter into proper fusion which is abated in slow degrees.

I am persuaded that if an exact knowledge [82] of the shape of the crystals of each metal, or rather of their molecules, were acquired, it would be possible to discover the kind of metal that would form alloys from the shape that would be taken by the crystals of several mixed metals that are capable of union; and the proportions of the mixture could be found by computing the number of angles and their opening as well as the greater or less extension of the faces of the hybrid crystals [*crystaux métis*] of the alloyed metals, for two bodies when united acquire together a mean configurational form that depends on their proportions.

This idea of mean relative proportions enabled the wise Archimedes, by means of the laws of hydrostatics, to discover the quality and quantity of the alloy by which a goldsmith had altered the gold in the crown of Hieron II. We know by inspection alone, by tasting, or even by touching, which base and which acid contribute to the formation of almost all neutral salts: if we do not have the same knowledge about metals, it is because we have had less experience.

I pass on to the destruction of iron, which I shall run through quickly. Of all metals iron is the one that maintains its state the least well. Its abundant earthy principle is but imperfectly united with its phlogiston on account of the considerable portion of saline substance in its essence, which is detected by the taste that it imparts to the tongue. This is why all fluids, except mercury, attack and corrode it, though to different degrees.

At the surface of iron air makes a thin layer of rust, which as it hardens gives the effect of a fused, impenetrable varnish resembling the precious varnish on ancient bronzes.

Water reacts on iron more actively than air because it has a great affinity with the saline parts of iron. It combines with them and, because [*the iron*] loses its phlogiston, the earthy molecules are released and condense to form an ochery mineral if it is in the open air or a black-colored mineral whenever this solution occurs under water. [83] The more sulphurous saline parts the iron contains, i.e., the more imperfect it is, the more easily does water destroy it.

All saline substances attack iron with ease and reduce it to rust more or less quickly. All kinds of iron rust are entirely analogous to ores formed by erosion or by deposition.

Sulphur has a very great affinity to iron, attacking it when cold if there is some moisture to facilitate movement; when hot, sulphur melts iron in an instant and reduces it to the state of pyrite.

Fire always acts upon iron. Whenever the iron contains superabundant sulphurous parts, the fire removes them and thereby perfects it. However, when it has acquired its degree of perfect metallization, fire attacks its own substance and destroys it. A vehement fire penetrates iron intimately, separating and dividing its molecules; the saline part of iron combines with its own phlogiston as well as with that of the charcoal. Sulphur forms, which melts and vitrifies the earthy part. What I advance is proven by a common experiment: if water is interposed between iron that is hot enough to be about to melt and the anvil, and if the iron is struck sharply with a hammer, there is a very violent explosion that disseminates a sulphurous odor: every molecule of water, very much divided, becomes charged with a molecule of artificial sulphur which follows it in its rapid course.

If the fire is less violent but continuing, it reduces the iron to a very attenuated red powder; it is properly a colcothar formed by the earthy principle of iron, greatly attenuated by the sulphurs, deprived of saline parts, and still charged with phlogiston. Vitrification did not take place because the fire was not strong enough.

I once had a vitrification of this colcothar made in a very violent fire that was slowly extinguished. It was crystallized in a regular shape and was of a ruby spinel color. Since it was apparent only to interested eyes, for it [84] lay on a very heavy crud, I was deprived of it as a result of [*a charwoman's*] excessive misguided cleanliness.

Finally, when iron is exposed to a less intense fire, it imperceptibly loses its active principles at the surface; the destroyed parts form leaflets that cover it and that multiply as they tend toward the center in proportion to the duration of the action. Iron in this state closely approaches that of the brown and hard bloodstones which are often very rich ores.

In general, however strong a fire may be, it cannot melt iron in its state of perfection without denaturing and mineralizing it. The spark that comes forth from a fire-steel is a small globule of iron that was detached by the shock, made red-hot and melted by the brisk friction to the point where it loses its phlogiston by contact with the air, with an explosion that makes this little bomb burst.

By continuity of action both fire and water destroy the aggregation of iron, though in different ways. The one removes, the other imparts, something. When water attacks iron, it first attaches itself to the weakest part and removes the saline and superabundant sulphurous part, so that if iron is exposed for a long time to the impression of water and is not totally destroyed, the portion that escaped disaster is a fibrous iron of a superior quality. On the contrary, fire attacks the true substance of iron immediately by inserting itself between its molecules; it distends them, inflates them, often vitrifies them in such a manner that the outline of a bar of iron that has been long exposed to great heat is surrounded by a hard and fragile, semivitrified crust similar to a slag. The interior parts that have not been completely destroyed were penetrated by the most active parts of the sulphur that formed and became so fragile that it is no longer of any use, having nothing but the appearance of a metal.

There are some ores of iron that rust and decay, [85] while others are perfected, depending upon the substances by which they are penetrated subsequent to their formation.

Cast iron is also susceptible to destruction. Water does not attack cast iron as easily as it does wrought iron, for we must still regard cast iron as being encumbered by a semivitrified material, and of all the states that bodies can have, vitrification is the one that renders them least accessible, because the vitrified parts are so attenuated and form so perfect a union that they become homogeneous and transparent.

The less perfect cast irons are, the less does water attack them. But fire eats into them all and destroys them more or less exactly depending on how it is applied. The first effect of fire on cast iron is to remove the superabundant sulphurs, bringing it that much closer to the state of a metal by making it reguline. In this consists the art of malleableizing cast iron and rendering it fileable; but if the fire is continued, the sulphurous principle that the fire develops is detached from the most interior parts of the cast iron, passes again through the parts located at the circumference, and, by a sort of palingenesis, it returns it to its primitive state. Alternately, the fire destroys and scorifies the iron by its superabundance and repeated action. I have a piece of cast iron that underwent these three degrees of alteration, whose center is gray cast iron surrounded by a bright, white, reguline band and enveloped in a black, hard, fragile crust made up of the debris of cast iron that has mineralized. (See Pl. VII, Fig. 21.)

Whenever the action of fire is aided by alternations with humidity, the destruction of the cast iron is quicker and more complete, because water insinuates itself into the pores that were opened up by the action of the fire, so making a wedge and lifting up the particles altered by fire wherever it exposes new surfaces. Moreover, the saline parts of the cast iron combine with the water to form a vehicle [86] that becomes more powerful as it becomes more similar and as the heat gives action to it.

Chimney backs in ground-level rooms that open on a yard or a street perish in a very short time because the humidity from the outside (often charged with nitrous or ammoniacal parts) is attracted by the fire inside and works in concert with the heat of the fireplace toward the destruction of the cast iron of the plaque. They reduce the plaque to a brown friable substance composed of layers applied one upon another, whose thickness is proportionate to the periods of action and which have a resinous appearance like those large red tartars from the Rhine.[13] Subterranean vapors destroy iron. I have a piece from the ramps of the cellars at the Observatory that has acquired ten times its [*original*] volume.

All the observations spread throughout this Memoir lead naturally to define what iron is: a metal of a loose texture, hard, and of limited ductility; gray in color, dark on the fracture, its polish resembling that of the water of a cloudy diamond; it attains its perfection only after it has passed through a number of different states, each of which has a quality that makes it suitable for the arts; [*it is a metal*] that easily receives heat by impulse and conserves it a long time; that requires the most vehement fire in its treatment; that expands considerably and softens in fire and contracts and

---

[13] The plates of cast iron that cover the vault under the hearth of a [*reverberatory*] furnace in a foundry are often destroyed in a year or two: they become more fragile than glass.

hardens by cold; that cannot be made white-hot without losing some of its substance; that never melts in the fire until after it has lost its state of ductility; that is dissolved and destroyed by all fluids except mercury; it is composed of an abundant vitrifiable earth, a sulphurous salt, and much poorly combined phlogiston, [87] affecting the form of a rhomb when all together, grouped differently according to their proportions when they have not yet acquired the degree of perfect metallization, whereafter they appear to be arranged in threads that seemed to me to be tubular, united in bundles within a common envelope.

The many kinds of recrement from iron forges that go indiscriminately by the name of slag differ greatly among themselves. Their analysis could spread light on iron and steel technology, as much on their nomenclature as on the knowledge of their physical causes, and could bring the manufacturing procedures to perfection.

EXPLANATION OF PLATE V

Figure 1. The segment of a cannon ball in which the center of crystallization has been displaced and pushed toward B because of an accidental cooling that occurred on the opposite side, along the line marked A.

Figure 2. Segment of a cannon ball of white cast iron in which crystallization occurred naturally in radii converging toward the center.

Figure 3. The shape of a crystal of white cast iron.

Figure 4. Segment of a cannon ball cast from tumultuously agitated cast iron, which produced some cavities, E, and others spread throughout the mass. These cavities decrease the weight without decreasing the volume. Part C condensed confusedly without arrangement, the empty spaces and the movement having disturbed the order and changed the natural direction at D.

Figure 5. Segment of a bomb with holes in the thickness [*of its walls*]. A thin layer of cast iron often covers the cavities, and if part F (the one that lies on the charge in the mortar) is weak, it is split by the force of the explosion so that fire is communicated to the powder in the interior and makes the bomb burst: if the cavities are numerous, the weight of the bomb is proportionately diminished, from whence arise errors in its [*trajectory during*] use. The handles, G, must be of very good wrought iron because cast iron conveys a brittle quality to iron.

Figure 6. Cube composed of small cubes (Fig. 7) piled up. If the base of a cube, which we can suppose to be of one twelfth of an inch across, is composed of a row of one thousand molecules of matter, each layer will contain one million; that will give a cube composed of a total of one thousand million molecules. Consequently, a cube of one inch in diameter will contain 1728 billion [1.728×10^{12}]. It is the same with the rhomb.

Figure 8. A rhomb having the shape of the molecules of iron, composed of an infinite number of small rhombs (Fig. 9).

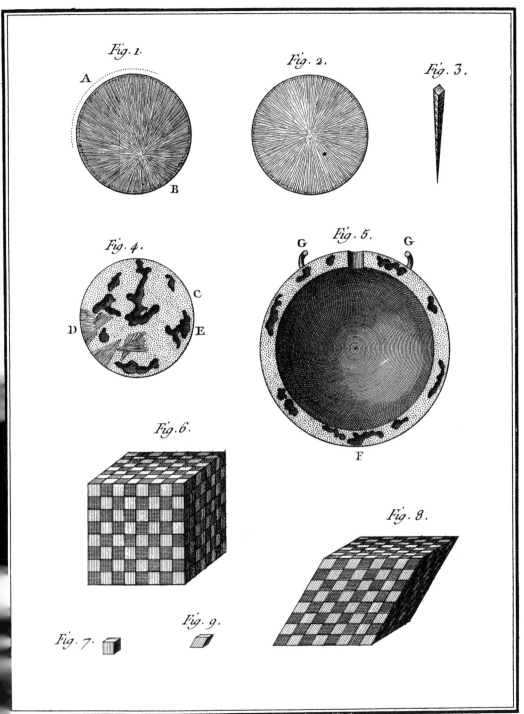

Plate V

EXPLANATION OF PLATE VI

Figure 10. A piece of cast iron, the surface of which is strewn with crystals of gray cast iron. These are scattered in a confused way and have mossy corners because the shrinkage was not great enough and they were brought too much to the surface.

Figure 11. Crystals of cast iron seen in their natural situation from different points of view and enlarged three times. [*In the original engraving these measure 1.05 to 1.4 cm., making the actual crystals 3.5 to 4.7 mm. long.*]

Figure 12. A crystal of gray cast iron, considerably enlarged and viewed from in front.

Figure 13. The same crystal viewed perpendicularly.

Figure 14. The same crystal viewed obliquely; the cruciform opposition [*of crystal branches*] shows at the base of the principal pyramid, L.

Figure 15. A crystal of a regulus of reguline iron, in the form of a rhomb.

Figure 16. A crystal of a regulus of iron, formed by the grouping of triangular prisms M N, which are sections of a rhomb.

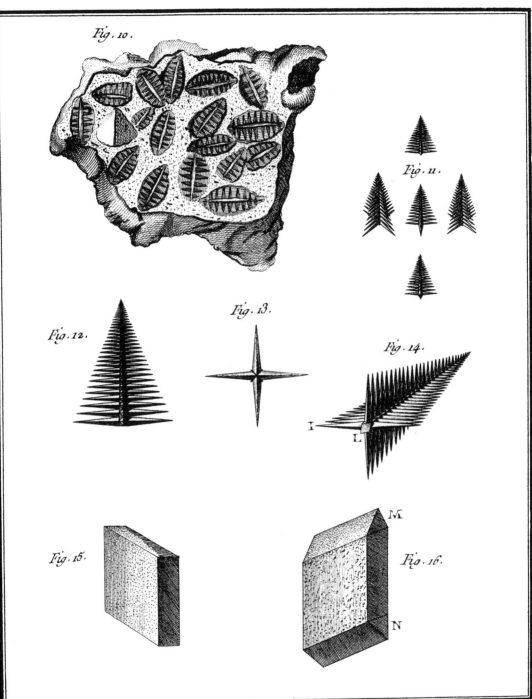

De la Gardette Sculp   Plate VI   Linéavit Grignon

EXPLANATION OF PLATE VII

Figure 17. A crystal of iron regulus forming a decatetrahedral-hexagon [*i.e., a 14-sided polyhedron based on 2 hexagons*]. Although the basis of iron crystals is a rhomb, it is not astonishing that they affect the polyhexahedron, which is simply several rhombs and sections of rhomb, as is shown in Figure 18.

Figure 18. Decatetrahexahedral-hexagon composed of rhombs and sections of rhombs. The trapezoid O, which diagonally crosses the upper segment of the hexahedron, is composed of two rhombs and a section of a rhomb, as is its opposite, Z. The triangles P and the others are so many sections of the rhomb, laid upon the first, and which, all being united by their rectangle surface, give a hexagonal polyhedron.

Figure 19. [*Text erroneously says 20.*] A piece of iron regulus wherein various rhombs, R, are oddly grouped. S presents a hexahedron; X, a triangular prism; O, platelets broken by main strength. Q is a confused crystallization because the matter has been less purified and rapidly cooled. T shows a tuft of asbestos [-*like fibrous crystals*] lodged within the thickness of the regulus. In the upper part several cavities, A, are seen, filled with ferruginous asbestos, separated by concentric strata, X. The partitions between the cavities are streaked with the impressions of asbestos, V. The top of some of these is cut into sharp teeth, Y; these teeth are so many crystal angles.

Figure 20. A group of little hexahedral crystals of reguline iron, shown considerably magnified in Figure 17.

Figure 21. Piece of cast iron whose first outer surface B has been destroyed and scorified by the continuity of the action [*of fire*]. The second layer, which is brilliant, is part of the cast iron that became reguline while the center retained the nature of cast iron, not having undergone the action of fire long enough to have passed to the state of regulus like the brilliant band that surrounds it nor to have been destroyed as was the external surface.

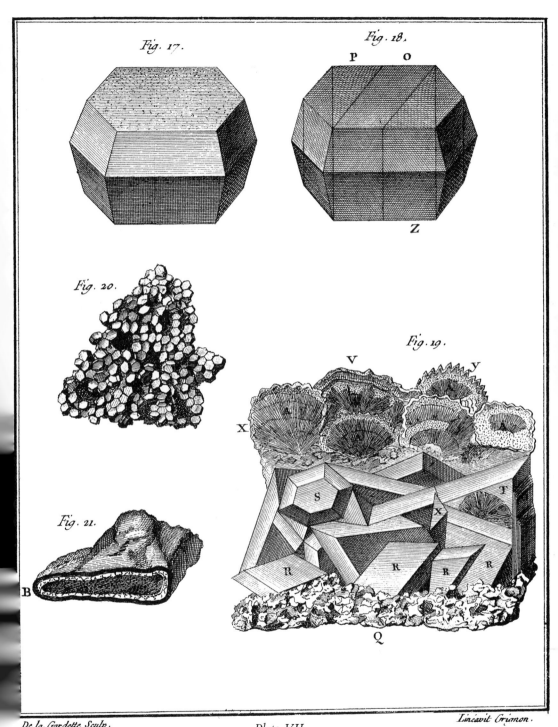

Plate VII

EXPLANATION OF PLATE VIII[j]

Figure 1 represents a very beautiful crystallization of cast iron. In the upper part A, A, A, and on the surface of the mass, as well as in the cavities, can be seen a great number of crystals, both grouped regularly together and simple like those shown in Plate VI. Other double compound crystals are recognizable, as in Figures C and D.

A, A, A. Regular grouped crystals of cast iron.

B. A cubic crystal of reguline iron, lodged in the artificial gangue, which adheres to the cast iron.

C. A double-compound crystal of cast iron. It is inclined in order to show the cross-shaped formation of the attachment of the grouped crystals on the principal and subordinate pyramids, which are all quadrangular both in their bases and in the disposition of the groups of additional crystals.

D. A double-compound crystal of cast iron. There are represented only two opposite parts of the principal column, with additional crystals implanted on the lateral columns. It can be seen that the subordinate pyramids in the center are the only ones that are furnished on both sides with additional crystals. The others, which continue to decrease up to the summit, are covered with these crystals only on the side toward the [*principal*] apex. Throughout the entire distribution of the different parts, all of the crystals can be seen to be in the form of diamonds in section and rhombs in mass, these figures, articulated one upon the other, being the elements of the crystals.

Figure 2 represents a vitreous crystallization of the ferruginous lava from a furnace smelting iron ores. These groups of crystals are implanted on a mass of artificial gangue, which contains a regulus of iron, crystallized in a cube, G,

---

[j] *This plate, included for its general interest, illustrates Grignon's essay entitled, "Mémoire sur des Crystallisations Métalliques, Pyriteuses et Vitreuses Artificielles, Formées par le Moyen du Feu," in the* Mémoires, *pp. 476–81. This was written apparently in 1773 or 1774 as an answer to Romé de l'Isle, who had contended in his* Essai de Crystallographie (Paris, 1772) *that water was the principal and perhaps the only medium of crystallization and that the feeble sketches of crystals produced in the laboratory should not be confused with the true crystallization forms produced by nature using the intermediate agency of water. Grignon claims that the ironmakers' furnaces are instruments with which art most closely approaches nature in imitation of volcanic products and that water had no part whatever in the formation of the crystals that he described. He even took a full-sized blast furnace and cooled it slowly over periods of fifteen to twenty days to get good crystals of metal and slag. He concludes that "It would be very interesting for the progress of physics if all men who employed fire on a large scale as an instrument in the arts could maintain during their work an observant analytical viewpoint. It would soon increase the extent of knowledge and diminish the relative number of aqueous systems."*

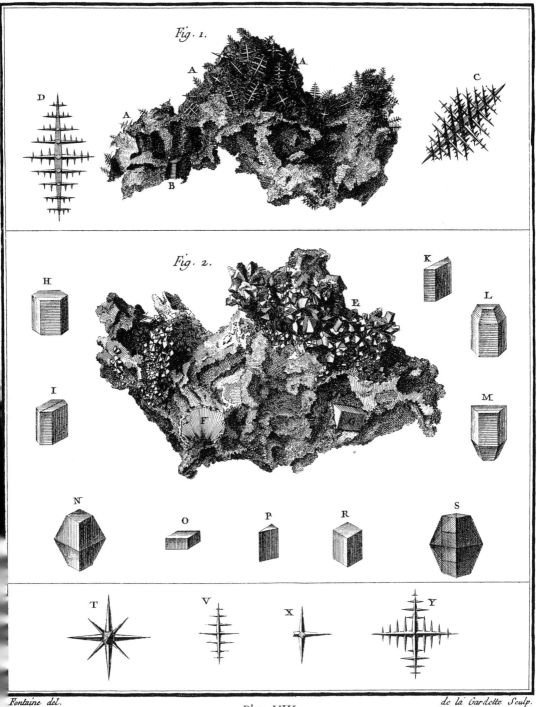

Plate VIII

and reduced to asbestos in F: the shape of vitreous crystals is developed in Figures H, I, K, L, M.

H. Regular vitreous crystal. It is a hexadral prism with four small and two large faces: it resembles crystals of topaz.

I. The same crystal, divided in the middle, which shows that it is composed of two tetrahedral crystals, K, the base of which is a trapezium.

K. Vitreous crystal whose shape is a tetrahedral prism with trapezoidal base: it is an element of the crystals H and I. [*The description of detailed geometry of* L-M *and of chafery-slag* N-S *is omitted in this translation, as it is obvious from the figures.*]

T. Octangular crystal of brass. The base of each column is an octagon, and other octahedral or tetrahedral columns are implanted on each face; the columns are grouped rather like those of cast iron.

V. A quadrangularly grouped crystal of brass, seen in profile.

X. Crystal of brass, grouped on only three faces of the principal column, and truncated on the other face. Only one element of the simple crystal can be seen, which is ordinarily an octahedron attached by its base to one of the faces of the principal pyramid.

Y. A double-compound crystal of brass.

Note. I have drawn the shapes of the isolated crystals considerably enlarged with a magnifying glass so as to make their form and their composition more easily seen. Only the pieces in Figures 1 and 2 have been drawn to natural size.

CHAPTER EIGHT

A Chemical Essay on the Analysis of Iron

Torbern Bergman

Dissertatio Chemica de Analysi Ferri (Uppsala, 1781). Translated by J. P. Hickey and C. S. Smith.

DISSERTATIO CHEMICA
DE
ANALYSI FERRI,

QUAM,

VENIA A PLISS. FACULT. PHIL S.,

PRÆSIDE
M*AG.* TORB. BERGMAN,

CHEMIÆ PROF. REG. ET ORD., NEC NON EQUITE AURATO
REGII ORDINIS DE WASA,

PUBLICE VENTILANDAM SISTIT
JOHANNES GADOLIN,
ABOA-FENNO.

IN AUDITORIO GUSTAVIANO MAJORI
D. 9 JUN. ANNO 1781.

UPSALIÆ,
APUD JOH. EDMAN, DIRECTOR. ET R. . ACAD. TYPOGR.

Plate IX
Title page of first edition, Uppsala, 1781.
(Reproduced from copy in National Medical Library, Bethesda, Maryland.)

Editor's Note

For a brief time toward the end of the eighteenth century, iron and steel attracted the attention of the leading physical scientists of the day. In the short span of years from 1773 to 1786, the strong light of the newly developing analytical chemistry (so different from the methods of the ancient assayers though owing much to them) resulted in the throwing out from iron of all the ancient mysterious combustible and sulphurous principles and the clear realization that the various forms of iron were, in the main, simply due to varying amounts of elemental carbon in the metal. It was no accident that this work began in Sweden and reached its apex in France, in the same environments in which the role of oxygen in combustion was elucidated, and the names of the people involved are those centrally involved in the revolution in chemistry itself.

English metallurgists at the time seem to have been too busy profiting from Henry Cort's invention to worry about science, and the most important papers on the discovery of carbon in steel have not been available in English until the present translations.

Torbern Bergman (1735–1784) of Sweden was the leading chemist of his day, though at the end of his life he was eclipsed by the younger Antoine Lavoisier (1743–1794).[a] *The majority of his works*

---

[a] *The always useful* History of Chemistry *by J. R. Partington excels itself in discussing the Swedish chemists in the period under discussion.* ("Chemistry in Scandinavia," A History of Chemistry, *Vol. II, pp. 159–236.) A complete list of Bergman's publications is given by B. Moström,* Torbern Bergman. A Bibliography of His Work *(Stockholm, 1957). A recent brief biography is given by J. A. Schufle in* Chymia, *1967, 12, 59–98.*

are essentially contributions to the techniques of analytical chemistry and the unraveling of the composition of minerals. In addition to papers on specific topics, he gave a complete scheme for analysis by both the new wet methods and blowpipe analysis, which Bergman had learned from J. Gahn and was the first to record in a comprehensive, orderly form. He wrote a treatise on crystallography employing a qualitative model of crystals formed from unit blocks and illustrated the relationship between different crystal faces. His comprehensive classification of minerals, Sciagraphia Regni Mineralis Secundum Principia Proxima Digesta (1782), and a history of chemistry (1782) should also be noted. He wrote a most important "Dissertation on Elective Attractions" or chemical affinities (1775). Many of his papers were collected in his Opuscula Physica et Chemica in three volumes (1779–1783, with later editions).

He was a man of profound insight and wide knowledge. From our present viewpoint his most interesting paper is the Dissertatio Chemica de Analysi Ferri, written in Latin in 1781. This work was first published as a pamphlet of seventy-four pages, preparatory to its defense in public by Johann Gadolin on 9 June 1781, as part of his doctoral exercises. Gadolin, who was just twenty-one years old at the time, was a student of Bergman's and supposedly did many of the experiments that are described in the essay, although both concept and writing seem to be those of the master. Gadolin went on to an eminent chemical career of his own and did much original work on heat and the rare earths.

The work was reprinted with a shorter title, De Analysi Ferri, and a few changes reflecting Bergman's more mature opinions, in Volume III of Bergman's Opuscula Physica et Chemica (Leipzig and Uppsala, 1783), pages 1–107.

The original pamphlet of 1781 was translated into French by Pierre C. Grignon, who published it as Analyse de Fer (Paris, 1783), with copious though not always perceptive notes of his own. It was this edition that was largely responsible for the interest *of leading French scientists in the problem of steel. (See Chapters Nine and Ten.) Though Bergman's essay on iron was translated into German in 1785 by Heinrich Tabor as the first paper in Volume III of the* Kleine Physische und Chemische Werke *(Frankfort, 1782–90), it has never before appeared in English, for only Volumes I, II, and IV (the latter posthumous) were included in the English edition published in 1784 and 1791.*

There are slight differences between the Latin of the first and second editions. In this translation, we have generally followed the first edition but have carefully noted wherever an emendation in the Opuscula *represents a significant change of viewpoint on the part of the author.*

The work described in this important essay was done in the years of chemical ferment immediately preceding the "Chemical Revolution" sparked by the work of Lavoisier and his associates in France. Oxygen had already been discovered, and Bergman understood its role in combustion and calcination; he still, however, continued to use the phlogiston theory to explain the results of his analytical experiments on iron, with results that are fascinating illustrations of eighteenth-century chemical thinking. At the time, Swedish chemists were leading the world in the development and use of new analytical techniques, which they used to uncover new elements and as in the case of the present paper to show that many long-known anomalies in the properties of materials were associated with some minor chemical constituent whose presence or role had not previously been suspected. Not only were the Swedish chemists active in their analyses but Swedish metallurgical industries were thriving. It was, indeed, the establishment of a new factory for the manufacture of gunbarrels emulating the famed Damascus steel barrels of the Near East that precipitated the whole chain of work that led to the discovery of carbon so elegantly proved in the present paper. The manager of a gunbarrel factory, Peter Wäsström, reported his process to the Academy of Sciences in 1773, and the description of the final etching of the welded barrels caught the attention of the eminent metallurgist, Sven Rinman, who himself thereafter did some etching experiments of his own. He clearly outlined the role of carbon for the first time in history in a paper, Rön om etsning på järn och stål, *read before the Academy in 1774.*[b] *Bergman took Rinman's bare concept that some material taken from charcoal was present in cast iron and steel, and by a series of very well-planned experiments he provides a quantitative proof of it. Moreover, as a believer in the phlogiston theory and with an interest in the absorption and evolution of heat in chemical reactions, he invokes rather hypothetical changes in the amounts of two kinds of phlogiston and caloric to explain his results in much the same way as the latest theoretical concepts have always been over-exploited (witness the liberal and often unnecessary use of dislocations and Fermi levels in today's metallurgical papers). It should be noticed that Bergman was using a rather specialized and short-lived form of the phlogiston theory, in which the distinction was made between reducing phlogiston — simply that which converts a calx into a metal and which can be measured by the amount of hydrogen that is evolved when the metal is dissolved in acid — and congealing phlogiston, which seems to be somewhat related to bound-in caloric matter. Bergman was close to a good theory of thermochemistry and clearly understood that there are definite amounts of heat involved in changes of state and in the formation of chemical compounds and that these can be differentially measured, though his experimental results were rather inaccurate.*

It is interesting to compare Bergman's approach with that of the great French scientist, Réaumur, half a century earlier in his L'Art de convertir le fer forgé en acier *(Paris, 1722). (See Chapter Five in the present book.) Réaumur was far more sensitive than was Bergman*

---

[b] *An English translation of this paper is given in full as Appendix A, pp. 249–55, in C. S. Smith,* A History of Metallography *(Chicago, 1960). The general history of the discovery of carbon in steel is discussed in a paper by the same author in* Technology and Culture, *1964, 5, 149–75.*

to changes of structure and physical changes generally, but Bergman tried to measure everything quantitatively and sought, as did all chemists for a century and a half following him, the main explanation of properties in terms of chemical composition.

A Chemical Essay on the Analysis of Iron

Those things that are lacking in this little work, may the ages supply.
QUINTILIAN

Section 1
The Varieties of Iron

IRON generally serves mankind in three states: cast iron (*tackjärn* in Swedish) is fitted to our use by smelting; wrought iron (*stångjärn*), having been further skillfully shaped under the hammer, takes on innumerable forms for our utilization; and finally iron under the name of steel, possessing extraordinary elasticity, hardness, durability, and other qualities surpassing the preceding types by a large measure, provides material suitable for wonderful mechanisms. But beyond this, in whatsoever state it is found, there are multitudinous sources of its mutations. With respect to cast iron, not only is there a manifold variety of ores that are worked but also different shapes and sizes of furnaces are employed with different qualities and amounts of the combustible materials that are necessary for nourishing the fire, as well as different characteristics of the bellows with respect to both the force of the blast of air and its direction and abundance. Changing only the proportion of ore in relation to the amount of charcoal regardless of the order in which the materials are charged [*into the furnace*] at the same time begets many distinct types, each with its own proper name. Thus we have cast iron that is saturated with phlogiston (*nödsatt*); cast iron endowed with a moderate amount of phlogiston (*fullsatt, lagom malmadt*); and finally cast iron sparsely furnished with phlogiston (*bårdsatt, såttjårn*). [2]° I lump together the remaining divisions, which are of more or less importance, for we now, as the saying goes, outstrip them in the race.

In order for it to be made malleable, cast iron must be fully treated in the fire once more and must be skillfully worked in order that the largest quantity of the best wrought iron may be acquired at minimum cost. Apart from the nature of the cast iron itself, the structure of the hearth, the quality and amount of the charcoal, and the inclination and efficacy of the blast are all influences that may cause many differences; and even if all these were held constant, the control of the very operation itself nevertheless begets not a few differences. And regarding methods of forging, there are two that now flourish among us, one which is usually called the Gallic [*Walloon*] method (*Vallonsmide*), the other the Germanic (*Tysksmide*). Occasionally also a combination is used that differs somewhat from either of the preceding two. The process by which in former times a very small amount of Osmund iron (*Osmundssmide*) was obtained and that which gave the great mass of iron called Buth, which was worked by rustic mountaineers (*Buth-smide*), have been overthrown, as it were, in our time; in like manner also the "hurry-up" method (*Rånn-smide*), so called on account of its speed, since wrought iron can be furnished on the very day on which the ores were broken up or on the day immediately following.

Now let us cease to marvel that the metal per se is especially inclined to changes, that it produces widely different degrees of hardness, ductility, tenacity, and elasticity. That difference is especially noteworthy whereby one kind of metal is capable of being extended when cold, but when it is red hot, it is broken (*rödbråckt*) [*i.e., red-short iron*]; while another kind, on the other hand, is obedient to the hammer when it has been properly heated but, when cold, cracks and is split (*kallbråckt*) [*cold-short iron*]; and

° *Numbers in brackets mark the beginning of the pages in the first Latin edition, 1781.*

there is a third kind that is found to be tractable in both states (*smidigt*).

The preparation of steel presents a new array of differences both with respect to the material to be converted (remembering that it shares all the variations discussed above) and with respect to the additional processing. [3] Indeed, steel is generally prepared either by smelting (*smeltstål*) or by cementation (*brånstål*), but there is such a variety of names applied to either of these types when obtained from different raw materials by different preparation and with different uses and properties that we have felt that it would be better to pass them over in silence lest the enumeration be even more in vain unless a description of the processes is added at the same time, for the bare and not infrequently absurd terminology that arises in common usage is of no value here where we touch only on the fringes of the variations and consider them as from a distance.

Indeed no other metal is so deceptive in its nature as iron, even notwithstanding the variety of the ores and ore-matrices of the other metals, and even the methods by which they are derived; and so many differences never appear, unless perchance a little more of a foreign metal may be present. Nevertheless, although it is often very difficult to do this, it is necessary to discover in this mutability the particular basis of the manifold uses of iron, which lies hidden. If its principal properties, namely hardness, tenacity, ductility, and elasticity, were always of the same intensity, that aptitude would arise only if its boundaries were narrowly defined; but as each of these qualities appears in very many different degrees, their different combinations open the door to a limitless area for sagacious and industrious investigation, so that it rightly can be said that polymorphous iron simultaneously contains the various qualities of very many metals.

[In order that the best result may be achieved in any application of steel that has been undertaken, a special predetermined balanced proportion of the four properties noted above is required, although for our present purposes these should necessarily be varied in almost innumerable ways. (*2nd ed.*)]

Because of this, iron frequently surpasses in value all noble things, even our beloved gold, and rightly so. When a pound of iron has been drawn out into wire, its value increases a hundred times over that of wrought iron. The same amount of iron used in the construction of watches is seventy thousand times more valuable, and in even more delicate works of art it exceeds the value of wrought iron 1,600,000 times.[1] [4]

Not only do civilized nations, who often prefer ornate and artistic work to utility, place such a high value on iron but also wild and barbarous peoples who lack iron commonly prefer it above everything else, occasionally even risking their lives to steal a nail or to take it by force, as many travelers' journals tell us.

Section 2
On the Causes of the Variations to be Investigated

The many variations seem to be derived from well-hidden causes, since although iron has been studied and handled almost from the beginning of the world, its intimate nature still remains very obscure.

The reasons for the mutations are generally sought either in a foreign admixture that can be absent without impairing the essence of the iron or in the varied proportion of the constituent principles of the metal itself.

The number of heterogeneous things that can be joined with our metal is very great, but it does not follow that each of them is therefore to be blamed. Only those that are frequently present in the ores are properly suspect. This applies to sulphur and its analogue plumbago, and, among the metals, arsenic, sometimes

---

[1] A lecture by Master Ekström before the Royal Academy of Science in Stockholm in 1750. [*Printed in* K. Vetenskaps Akademiens Handlingar (*Stockholm*), *1750.*]

zinc, and perhaps always manganese.[d] It follows that the first thing to be investigated by suitable experiments is whether one or more of these materials is in fact present, then to inquire into what effect they may be able to produce.

That which pertains to the intimate composition of iron has not yet been laid bare, but in order that the way may be laid open to new attempts, let us pursue a sound analogy. We know that arsenic consists of a particular acid because a certain amount of phlogiston is condensed in white arsenic, but this coagulant, which is united to a definite portion of the inflammable matter, has departed in the complete [5] semimetal.[e] This connection now has been placed beyond all chance of doubt by experiments involving both analysis and synthesis, though it is indeed not permissible to conclude therefrom that every metal has been uniquely composed of an acid root [*acidis radicalibus*] and phlogiston or, more accurately, that their calces are particular acids that have been congealed by phlogiston. This is indeed a slippery method of reasoning, which

---

[d] *The Latin is* magnesia, *and the word "magnesia" was commonly used in contemporary English, including the translations of Bergman's other works. Black manganese and white magnesium compounds were long thought to have some relationship, but there is no doubt whatever that the author means manganese, and it seems best to avoid the confusion inherent in the old terminology.*

[e] *Bergman's form of the phlogiston theory had the complicatedness which often marks a theory that has run its time and is about to be replaced. In addition to the more conventional "reducing" phlogiston that combined with calces and thereby endowed them with fully metallic nature, Bergman also believed in a "congealing" phlogiston that was involved in the constitution of the calces themselves, for these were in general made of some acid congealed by its aid. Congealing phlogiston could be removed only with difficulty from the calces (except in the case of white arsenic), but the other kind was extractable with an ease that increased the more noble metal. The reducing phlogiston could be liberated from the metal by the action of acids, and it then formed fluids — inflammable gas (hydrogen) in hydrochloric or sulphuric acid, or a nitrous gas with nitric acid. These gases were measurable and could be used as a quantitative index of the amount of phlogiston in the dissolved metal. Alternatively (in a manner that can be regarded as foreshadowing Faraday's measurement of electrochemical equivalents), the phlogiston in iron could be used to precipitate silver quantitatively from silver sulphate solution or to reduce lead from litharge.*

should therefore be followed with great care (in a manner not unlike Ariadne's thread in the story) through the obscure passages of nature until it can be proven by carefully chosen experiments or our conjectures destroyed. Nevertheless in the present case it is a notable probability. When they lose their phlogiston, all metals go into powders that resemble earths, fusible and very heavy; and, if the phlogiston is restored, they are again endowed with a completely metallic nature. Might one not, therefore, logically suspect a similarity in the composition of these calces? If the phlogiston remains the same in all, but saturates different acids, it would explain the specific differences between the metals. Nothing so far is inconsistent with the phenomena noted; indeed there is much that supports it. The fact that the spagyric art has not yet been able to discover any root [*of an acid*] except arsenic does not hinder [*the argument*], for the force by which the congealing phlogiston is retained can be so great that those mediums that have been applied have been unable to separate it even to this day, and it may be that an effect equivalent to this will remain hidden through the ages. That portion of phlogiston that reduces metallic calx into its complete form is usually easily separated; however, its association is nevertheless so unequal as to form the basis of the division of metals into noble ones and base. If the difference is so conspicuous in the behavior of the reducing part [*of the phlogiston*], why can it not also be in the congealing portion?

Let us now proceed to apply this theory, which we now recognize as highly probable, to iron, in order that through new experiments it may afford the opportunity for the hypothesis to be either confirmed or else utterly destroyed. The first question of all that may be put regarding [6] reducing phlogiston is: "Is it in fact possible to have any variation in the quantity of it in iron?" If such a variation can exist, then indeed its effects in various instances are to be investigated. Once its limits have been set, the examination may then proceed to the congealing phlogiston and how it may be tested if it is going to be taken away, so that it may afford an acid root stripped bare by itself for a separate careful investigation into its abundance and properties.

Moreover, some amount of bound caloric matter is always present by chance in all bodies, a thing that can appear conspicuously, especially in metals. But whether the proportion differs perceptibly in the different states of iron, and what the effect of this variation may be, is to be determined by experiments.

We have expanded this outline appropriately in a special section in which we report many experiments. Although many remain still to be zealously pursued, it will not, perhaps, be in vain to divulge these fragments in the meantime, despite the fact that they are still imperfect and incomplete, for we hope that others will be stimulated by them in this task even though they may lack at the time the most useful facilities. It is necessary that the space of many years be devoted singlemindedly to this labor, and indeed, in addition to the necessary apparatus, mental ability is required not only to conceive suitable tests that will throw light on the subject but also to arrange carefully experimental methods that are appropriate for realization. If ever any influences are to be recognized with certainty, it calls for unlimited tirelessness, turning over in the mind all the phenomena that have been correctly determined and drawing from them careful conclusions.

Section 3
The Quantity of Reducing Phlogiston Sought by Experiments in the Humid Way

Granting that it may be apparent from the preceding discussion, we add here *expressis verbis* a useful warning in order to avoid ambiguity: we wish it understood that by reducing phlogiston we mean [7] that portion of phlogiston the removal of which causes a metal to lose its complete form and to become dissolvable in acids without the production of any aeriform fluid and the restoration

THE ANALYSIS OF IRON 179

of which causes the retrieval of its completely metallic nature. The whole of the inflammable matter that remains within we attribute to the congealing portion, unless any foreign matter (the offspring of the same hidden principle) is present at the same time.

In order that we may understand more accurately this abundance of phlogiston, we have not been content with a single approach but have at the same time tried another different method leading to the same end. Let us consider the basis of the first method. It has been noted[2] that when metals are treated with acids they give off reducing phlogiston, which on solution generates aeriform fluids that are different according to the circumstances. Vitriolic and muriatic acids with iron produce the air that is commonly called inflammable air, but nitric acid[f] brings forth [*nitrous*] air, so named from niter. Since these fluids contain phlogiston that enters into their composition, we have decided that the volumes of these fluids that are generated are proportionate to the quantities of inflammable air released, a thing that we have found consonant with the experiments on inflammable air soon to be noted.

(A) The tests were set up in the following manner. A glass flask was used, having a narrow neck at the top, to the upper extremity of which an ∽-shaped tube had been fitted by grinding, so that when in place it sealed perfectly all the way around. Thus this flask could be placed in a copper vessel, the cover of which had a hole that served to support the neck securely, while the tube could very freely enter the opening of an inverted earthenware jar filled with water and suspended at an appropriate height with its mouth immersed beneath the surface of water in a vessel placed under it. Things having been thus prepared, water was poured into the copper vessel and, with fire, brought to a boil. Meanwhile a

---

[2] *Opuscula*, Vol. II, p. 354.
[f] Acidum nitrosum: *We have used the modern term "nitric" acid in preference to the contemporary "nitrous" to avoid confusion. It should be unnecessary to remind the reader that vitriolic and muriatic acids are, respectively, sulphuric and hydrochloric acid. Hydrochloric acid was also "acid of salt."*

hundredweight of iron[g] reduced to powder by filing or hammering, depending on its nature, was poured into the flask, the neck of the flask [8] was secured in the opening of the cover, the necessary volume of acid solution was added to the iron, the bent tube then inserted with its free end remaining in the mouth of the earthenware jar, and finally the flask was carefully immersed in the water in the copper vessel and the cover put in place on the copper. When this had been done, the water was brought to a boil by the fire and the boiling specially maintained as long as any amount of air escaped into the earthenware jar so that the water might not fill up the globe as it would if by chance the heat should diminish. In all the experiments an equal volume was taken from the same bottle of vitriolic acid, as was also done with the other acids that were added in the subsequent variations. Therefore, since the same apparatus was used in all experiments and the operation was aided by the same degree of heat, the circumstances were the same, and only the quantity of the iron was varied, as was suitable. We measured the aeriform fluid in tenths of a cubic inch, each of which contained 2 loths [or, if you will, 800 assay pounds (*2nd ed.*)] of pure water.

The flask with the tube held $8\frac{1}{2}$ loths of water, i.e., $4\frac{1}{4}$ cubic inches, and when it was immersed dry in boiling water, only $1\frac{1}{4}$ cubic inches of the enclosed air expanded into the earthenware jar. When the same test was made with 2 loths of water (approximately corresponding to the amount of the solution used in the later experiments), the force of the vapor expelled 3 [*cubic*] inches of air into the earthenware jar, which therefore must be subtracted

---

[g] *Throughout this essay the weight system is, of course, the miniature system of assay weights, not legal pounds. According to the statement later in this paragraph, 800 assay pounds are equivalent to one ounce, which makes the hundredweight equal to one Swedish kvintin or 3.22 grams. Grignon in his translation assumes that the assay hundredweight is 100 Swedish grains, which he says are equivalent to $90\frac{2}{3}$ French grains, which would be 4.84 grams in the later metric system. Grignon makes another error in giving the Swedish cubic inch as equal to 1.2025 that of Paris; it was actually smaller. The Swedish inch (tum) was 2.474 cm.*

from the total volume collected. This has already been done in the tables following.

(B) We first tried vitriolic acid. That which is mentioned in our experiments had been purchased under the name of oil of vitriol and diluted by four times its weight of distilled water. Its specific gravity was 1.129.[h]

The first column of the attached table records the number of cubic inches of air extracted from the hundredweight of iron, properly corrected by subtraction [*as just noted*].

The second column indicates the duration, in minutes, of the time of collection up to the point where no more air was being produced. [9]

[Experiment no.]	[Kind of iron]	[Gas evolved] cubic inches	[Time of solution] minutes
Expt. 1.	Cast iron saturated with phlogiston, taken from the great smelting furnace at Leufstad in Roslagia, gave	43	45
Expt. 2.	Cast iron endowed with a moderate amount of phlogiston, from the same furnace at Leufstad	39.5	45
Expt. 3.	Wrought iron from the same place at Leufstad	50	15
Expt. 4.	Ductile iron from the same cast iron prepared in the crucible (Expt. 94)	51	25
Expt. 5.	Cast iron coming from Åkerby	38	50
Expt. 6.	Wrought iron from Åkerby, as the preceding	48	15
Expt. 7.	Cast iron from Ullfors	41	45
Expt. 8.	Wrought iron from Ullfors	50	15
Expt. 9.	The same, with a double quantity of vitriolic acid	50	15

[h] *The specific gravity of 1.129 corresponds to about 19 per cent sulphuric acid according to today's tables.*

[Experiment no.]	[Kind of iron]	[Gas evolved] cubic inches	[Time of solution] minutes
Expt. 10.	The same dissolved simply in the heat of digestion	50	—
Expt. 11.	Wrought iron from Österby	48	15
Expt. 12.	Steel made by the cementation of this iron, hardened, but softened in the fire so that it could be filed	46	10
Expt. 13.	The same steel heated, extended into sheets under the hammer, quenched to make it hard, and powdered	46	30

[10] The varieties of iron so far enumerated [*Experiments 1–13*] come from the surface ores at Dannemora, and had been smelted and wrought by the Gallic [*Walloon*] method. These varieties are especially praised by foreigners under the name of *Oregrundian iron*.[3]

[Experiment no.]	[Kind of iron]	[Gas evolved] cubic inches	[Time of solution] minutes
Expt. 14.	Cast iron from Forsmark	40	55
Expt. 15.	Wrought iron from Forsmark	51	15
Expt. 16.	Steel coming from fagoting the same iron, hardening it, and then heating so that it could be filed	48	10
Expt. 17.	Cast iron from Brattefors in Värmland	41	20
Expt. 18.	Wrought iron made in the German fashion from Brattefors	51	10
Expt. 19.	Native iron from Siberia [*i.e.*, a meteorite]	36	240

[3] The iron smelted from Dannemora ores was formerly exported through Oregrund, and, although the town lost the benefit of this trade as long ago as the year 1638, the iron is nevertheless, even to this day, referred to by that name. [*In the first edition the interpolation on Dannemora iron is erroneously placed after Expt. 16.*]

THE ANALYSIS OF IRON

[Experiment no.]	[Kind of iron]	[Gas evolved] cubic inches	[Time of solution] minutes
Expt. 20.	Cast iron surcharged with phlogiston from Hällefors in Sudermania scraped from the middle by the boring mill	48	15
Expt. 21.	The same remelted in a crucible	43	45
Expt. 22.	English steel produced by cementation and then melting; of unimpaired ductility	45	12
Expt. 23.	Steel prepared in the same way by Dr. Quist	46	6
Expt. 24.	Cast iron from the ores encountered at Dingelvik in Dalia, which are contaminated with manganese	41	90
Expt. 25.	Steel prepared by cementation of wrought iron from the preceding cast iron	47	15
Expt. 26.	Wrought iron from the parish of Norrberg, hot short	48	10
Expt. 27.	Wrought iron from the parish of Gränjen, cold short	51	8
Expt. 28.	Cast iron coming from Husaby in Smolandia	48	30
[11] Expt. 29.	Wrought iron from the preceding, cold short	50	6
Expt. 30.	Steel prepared by cementation of the preceding	44	25
Expt. 31.	Cold-short iron coming from Braås in Smolandia, from a small Hungarian furnace	52	25
Expt. 32.	The same as the preceding rendered malleable with the aid of limestone from Braås[4]	48	20

[4] The assessor of the Royal College of Metallurgy, the honorable A. Stockenström,[1] has communicated to us an account of some notable experiments carried out last

[1] *This is the same man who on 13 November 1781 wrote a letter to P. C. Grignon in France conveying an enthusiastic report of Bergman's freshly published paper. Grignon was eventually able to borrow a copy that had been sent to the Académie des Sciences and was in the possession of Berthollet, and he used it in making the French translation that was published in 1783.*

[Experiment no.]	[Kind of iron]	[Gas evolved] cubic inches	[Time of solution] minutes
Expt. 33.	Wrought iron remelted in a crucible with iron calx, but rendered no less cold short thereby (Expt. 100)	52	20

(C) Many of these experiments have been repeated using muriatic acid, the specific gravity of which was 1.155.[j] Several new specimens were added, especially the calces that are listed. [12] A hundredweight of iron was employed for these tests, as for those in the preceding table, and the same was done with the material that follows under section D.

[Experiment no.]	[Kind of iron]	[Gas evolved] cubic inches	[Time of solution] minutes
Expt. 34.	Cast iron from Leufstad saturated with phlogiston yielded	43	40
Expt. 35.	Cast iron from Leufstad with a moderate amount of phlogiston	39.5	30
Expt. 36.	Wrought iron from Leufstad	50	10

---

summer at Braås in Smolandia. Some ores, namely the bog ores, yielded only cold-short iron in the great smelting furnaces there. Doctor Stockenström, hoping to be able to make it malleable by a simple smelting operation, prepared a furnace, 8 feet high, like the kind that they call in Hungary, *Blau-ugn*. After repeated tests in this, he found that (1) when the ores were smelted without any addition, only cold-short iron was produced, which was the kind examined in Expt. 31; (2) when ⅛ of grey limestone contaminated with quartz was added, the ore produced malleable iron (studied in Expt. 32); (3) when the scale [*slag, in 2nd ed.*] that falls off during hammering (*Hammarslagg*) had been smelted and reduced, it also afforded cold-short iron; (4) a great loss of weight occurred in all smelting operations, whether or not limestone was used. The loss occasionally reached 50 per cent when the ore was charged, and when iron scale was employed it reached 28 per cent, although a hundredweight of ore lost only 30 per cent on calcination. It is obvious that both these amounts that are charged and those that are extracted should be weighed, as this demonstrates. (5) From a hundredweight of ore that promised about 24 pounds of cold-short iron, a total of [*only*] 18 pounds of ductile iron was obtained.

[j] *This specific gravity corresponds to about 30.5 per cent HCl.*

[Experiment no.]	[Kind of iron]	[Gas evolved] cubic inches	[Time of solution] minutes
Expt. 37.	Cast iron from Åkerby	38	30
Expt. 38.	Wrought iron from Åkerby	48	10
Expt. 39.	Cast iron from Ullfors	41	25
Expt. 40.	Wrought iron from Ullfors	50	10
Expt. 41.	Wrought iron from Österby	48	10
Expt. 42.	Steel from Österby, hardened	46	20
Expt. 43.	The same softened in the fire	46	5
Expt. 44.	Cast iron from Forsmark	40	25
Expt. 45.	Wrought iron from Forsmark	51	10
Expt. 46.	Steel from Forsmark	48	5
Expt. 47.	Cast iron from Brattefors	41	10
Expt. 48.	Wrought iron from Brattefors	51	10
Expt. 49.	Native Siberian iron	49	70
Expt. 50.	Cast iron from Hällefors, saturated with phlogiston, removed from the middle on the boring machine	48	10
Expt. 51.	The same remelted in a crucible	43	45
Expt. 52.	English steel	45	5
Expt. 53.	Steel made by Dr. Quist	46	4
Expt. 54.	Steel made from Dingelvik iron	47	10
Expt. 55.	Wrought iron from Norrberg	48	10
Expt. 56.	Wrought iron from Gränjen	51	4
Expt. 57.	Cast iron from Husaby	48	45
Expt. 58.	Wrought iron from Husaby	50	4
Expt. 59. [13]	Steel from Husaby iron	44	20
Expt. 60.	Cold-short iron from Braås	52	4
Expt. 61.	Ductile iron from Braås	48	8
Expt. 62.	Ductile iron melted with charcoal powder (Expt. 107)	45	45
Expt. 63.	Cast iron melted with glass (Expt. 115)	45	50
Expt. 64.	Cast iron melted with calcined lead (Expt. 110)	43	120

(D) We have also carried out experiments in the same manner using nitric acid, the specific gravity of which was 1.230.[k]

[k] *Acid of this specific gravity contains 36 per cent HNO_3.*

[Experiment no.]	[Kind of iron]	[Gas evolved] cubic inches
Expt. 65.	Cast iron from Leufstad, saturated with phlogiston, yielded of nitrous air	33
Expt. 66.	Cast iron from Leufstad with a moderate amount of phlogiston gave	29
Expt. 67.	Wrought iron from Leufstad	28
Expt. 68.	Cast iron from Ullfors	30
Expt. 69.	Wrought iron from Ullfors	30
Expt. 70.	Cast iron from Åkerby	26
Expt. 71.	Wrought iron from Åkerby. A little lump of 100 pounds placed in a lukewarm flask and then heated in water to boiling	29
Expt. 72.	The same in the form of filings; the flask immersed in somewhat hotter water	20
Expt. 73.	Filings of the same iron; set in boiling water	15
Expt. 74.	Wrought iron from Österby, in filings; the flask set in boiling water	18
Expt. 75.	Wrought iron from Österby in little lumps in the flask set in lukewarm water	24
[14]		
Expt. 76.	Cast iron from Hällefors, saturated with phlogiston; the flask immersed in water which was no more than lukewarm, then gradually raised to boiling	33
Expt. 77.	The same; the flask placed in somewhat hotter water	30
Expt. 78.	The same; while the globe was immersed in boiling water	28
Expt. 79.	Cast iron from Hällefors, poor in phlogiston; the flask immersed when tepid	34
Expt. 80.	The same; when the globe was immersed in somewhat warmer water	32
Expt. 81.	Hot-short wrought iron from Norrberg in filings; when the flask was immersed in hot water	19
Expt. 82.	The same; flask immersed in boiling water	14
Expt. 83.	Cast [sic] iron from Husaby, which was cold short	34
Expt. 84.	Steel made from the preceding wrought iron; the flask immersed in tepid water	32
Expt. 85.	The same, immersed in somewhat hotter water	30

(E) Now we shall also try to determine the quantity of reducing phlogiston in the different states of iron by precipitation. The obstacles that one encounters with nitric acid have been noted elsewhere.[5] Also, the amount of phlogiston in iron that had been deduced there on the basis of a single experiment with acid of vitriol was extremely uncertain, and this must be corrected. We have therefore attacked the matter in the following manner. Impure silver was dissolved in nitric acid to saturation, and the solution was precipitated by a vitriolated vegetable alkali.[1] A hundredweight of the [*silver*] vitriol after drying out contained 66.7 assay pounds of silver, a result that was determined both by precipitation and by reduction. [15]

Expt. 86. A hundredweight of vitriolated silver was put into about 6 [*cubic*] inches of distilled water, together with a few drops of vitriolic acid, not in order to promote the precipitation thereby, for this can be accomplished without them, but so that the deposition of a ferruginous powder might be impeded. Finally, a small well-cleaned lump of Österby wrought iron weighing 130 pounds was added, and the little vessel was exposed to a digesting heat until the liquid produced no turbidity with a drop of brine. When this condition had been attained, the small piece of iron was taken out, washed, and dried, when it was found to have lost about 19.5 pounds. It follows, therefore, that 66.7 pounds of silver were precipitated by 19.5 pounds of Österby iron.

Expt. 87. The test was repeated with another hundredweight of vitriolated silver, but in place of the Österby iron some Gränjen was used, of which just 17.9 pounds were sufficient for the precipitation.

---

[5] *De diversa phlogisti quantitate in metallis.* ["*Dissertation on the different quantity of phlogiston in metals.*" First published in 1782 and reprinted in the Opuscula, *Vol. III*.]

[1] *This treatment with potassium sulphate precipitates silver sulphate. When pure this contains 69.2 per cent of silver, compared with the 66.7 per cent reported by Bergman. In the next paragraph he proceeds to determine what amounts to the electrochemical equivalents of various grades of iron in displacing silver and regards the results as an index of the amount of reducing phlogiston present.*

Expt. 88. A hundredweight of vitriolated silver, when reduced by the action of Österby steel, required 20 pounds.

Expt. 89. A hundredweight of vitriolated silver, when precipitated by Husaby cast iron, had need for 19.2 pounds.

Section 4
Corollaries of the Preceding Experiments

If, now that the experiments have been carried out, they are compared and studied a little more carefully, it follows without difficulty that:

(A) Wholly equal volumes of inflammable air come forth from equal weights of the same iron with both vitriolic and muriatic acids, but there is a great discrepancy of time, for the first acid generally acts much more slowly; indeed, it often requires more than twice as much time. We believe that this discrepancy arises partly from the unequal states of division of the bodies to be dissolved, partly from the previous dilution of the menstruum. Nevertheless [16] either acid releases the same amount of inflammable air [*follows 2nd ed.*] whatever quantity of the menstruum may have been used, provided that there was enough (Experiments 8 and 9), undoubtedly because it is the phlogiston [*2nd ed.*] that particularly conveys the signal for the generation of this air. A peculiar property is possessed by the native iron from Siberia,[m] which causes it to be taken up very slowly by the menstruum. Even when helped by the heat of boiling water, vitriolic acid is able to extract only 36 cubic inches in four hours. Muriatic acid,

---

[m] *The Siberian native iron is, of course, a meteorite, undoubtedly the one found by Pallas, who had liberally distributed fragments of it to scientists throughout Europe. Bergman describes his examination of it in detail in his* Opuscula, *Vol. II, English translation, pp. 447–48. He concludes that it had been fused but that the stony matter was of a totally different nature from the slags produced in metallurgical furnaces. He found the iron to be tenacious and malleable except when it was strongly heated and, as reported later in this paragraph, sometimes to contain sulphur. This discovery of Bergman's is not noted in Harrison Brown's comprehensive* Bibliography on Meteorites (*Chicago, 1953*).

however, acts more quickly in the beginning, for in the first 10 minutes it produced 30 cubic inches, though in the following 60 minutes only 19. We have also observed such sluggishness in certain other natural products. Perhaps different pieces of Siberian iron differ in this regard, since we have not seen this behavior noted by anyone previously. In the specimen collection of the Academy at Uppsala there is one piece that gives rise to a strong hepatic odor with acid of salt, although the others afford no trace of this.

With regard to nitric [*acid*], the manner of its action depends on very minor circumstances, so that only very rarely can the same result be produced by an experiment that is repeated twice in the same apparatus with the same materials and quantities. Since both vitriolic and muriatic acids generate the same volume of air, regardless of whether they are aided by a boiling heat or by one of digestion (Experiments 8 and 10), we at first expected the same result with nitric acid, but we have observed that the production of air had been diminished not only with a more intense heat of the water but even when there was a greater surface for dissolving (Experiments 71–85). This makes us wonder even more how much more these factors may commonly augment the effects in other solutions. Furthermore, it follows simply from the experiments that have been carried out that this acid has been found to be less suitable for what we have proposed, and its effects must be examined in another place.

(B) Reducing phlogiston is proportionate to the volume of the inflammable air extracted: this is the foundation of the first method of measurement employed by us. In order that this hypothesis might be examined, [17] we have also measured the precipitation [*of silver*] produced by different varieties of iron, and we have found agreement, for 66.7 pounds of silver were reduced by 19.5 pounds of Österby iron (Experiment 86), though only 17.9 pounds of Gränjen iron were sufficient (Experiment 87). If from these facts the quantity of phlogiston in a hundredweight of each type of iron was computed, taking as basis that a hundredweight of silver contains 100, it would be found to be $\left[\dfrac{66.7 \times 100}{19.5} = \right]^n$ 342 for Österby iron, and $\dfrac{66.7 \times 100}{17.9} = 373$ for Gränjen iron. The former, however, yields 48 cubic inches of inflammable air (Experiment 11) and the latter 51 (Experiment 27), from which it ought to be that 48:51::342:373, which indeed is found to be close to the truth, for the fourth figure is actually 363, which lacks ten parts, although it would lack only one if the latter ratio were expressed in tens like the former [*i.e., to two significant figures*].

The following two experiments coincide similarly. In a hundredweight of Husaby cast iron, 347 [*units of phlogiston*] are present (Experiment 89) and in Österby steel, 333 (Experiment 88), but the former gave 48 cubic inches of inflammable air (Experiment 28) and the latter, 46 (Experiment 12). Therefore, it should be true that 48:46::347:333, but actually the fourth term of the proportion is 332.5, which differs by not even a single integer, and greater harmony among diverse methods is scarcely to be expected.

And so from this we conclude that there is present in one cubic inch of inflammable air about the same amount of phlogiston as in 2.17 pounds of Österby steel, 2.08 pounds of Husaby cast iron, 2.08 of Österby wrought iron, and 1.96 pounds of Gränjen wrought iron.

(C) Cast iron taken from the same ores contains different proportions of inflammable air, namely by an amount that is proportionately smaller as the charcoal had been added more sparingly during smelting. This is shown by many experiments that appear in Tables 1 and 2.

(D) The lowest limit of the variation in good cast iron is 38 [*cubic inches of inflammable air*], and the highest is 48, but we must frankly confess that the highest point seems [18] to us very un-

---

n *The part of the equation in brackets appears in the second edition only.*

certain. We have obtained this high value only with chips from Hällefors cast iron that had been scraped from the bore of the cannon at Hällefors and which had undoubtedly been contaminated by steel filings during the boring operation.° If this experiment is excluded as ambiguous, the highest value arising among the remaining seven is 43.[6] The cast iron from which cold-short wrought iron is produced does indeed evolve 48 cubic inches [*Experiment 28*], but we have not yet been able to examine the iron from Hällefors that gives wrought iron lacking this defect.

(E) The variation in the classes of steel runs between 45 and 48 cubic inches. That which had been prepared from cold-short iron drops to 44.

(F) The variation of good wrought iron falls between 48 and 51. We have been able to acquire only one specimen of hot-short iron. When this was investigated in the manner described, it stopped at about 48, the lower limit of good wrought iron, although we had three varieties of cold-short iron, two of which reached the high point of good wrought iron, namely 50 and 51, and the third even surpassed it, going all the way up to 52, as in Expt. 31.

(G) From the three sections immediately preceding it, it can be

[6] Unadulterated Hällefors cast iron was later acquired, and this gave 40 cubic inches in the accustomed method from a hundredweight, so the conjecture in the text regarding particles of steel having been mixed in is now confirmed. [*Footnote in second edition only.*]

° *By the time Bergman was writing, French and English guns were being bored from the solid, but the Swedish factories were still casting cannon with cores, and Bergman's sample would therefore represent only surface metal. There is a mention of a gun foundry near the town Nyköping in Sudermania in Gabriel Jars'* Voyages Métallurgiques *(Lyons, 1774), Vol. I, pp. 154–55. Jars says it is identical with another at Moss which he describes in more detail (ibid, pp. 172–73). The cannon were cast "in such a way that there remained only two or three lignes to be bored out." The boring machine worked perpendicularly, the cannon being pressed by its own weight on the boring tool which was driven by a water wheel and gears, evidently in a manner like that first depicted in 1603 (Prado y Tovar in A. R. Hall,* Ballistics in the Seventeenth Century *[Cambridge, 1952], Plate I) and shown in detail in Diderot's* Encyclopédie.

seen that the least amount of phlogiston is present in cast iron, the median amount is in steel, and the greatest amount is in wrought iron. But before we can regard this conclusion as altogether certain, it is necessary to know whether in these different states any foreign matter is present, either sparsely or in abundance, that releases more or less inflammable air, or none at all. This matter will eventually become known in Section 8. For the moment it may suffice to indicate briefly that, as is always the case where one speaks about the quantity of phlogiston in iron, only that phlogiston is to be considered that is proper to the iron molecules, for the part of the inflammable principle that is contained in plumbago and in other foreign materials can be absent with its own base, while the nature of the metal is preserved. [19]

Section 5
The Quantity of Reducing Phlogiston Sought by Experiments in the Dry Way

Many experiments were set up so that the amount of phlogiston might, if possible, be noted by this method also, but we have not yet been able to devise a method for exactly measuring it and have been forced therefore to use some kind of determination that tells in general whether a greater or smaller amount is present [*in a sample of iron*]. We have set out some typical experiments, arranged according to the diversity of the operations, though excluding not a few that seemed to us to settle little or nothing.

(A) The following experiments were carried out by melting and were actually done at the beginning [*of our researches*], in order that we might learn whether cast iron bears more phlogiston than does wrought iron.

Expt. 90. Two hundred pounds of Hällefors cast iron, saturated with phlogiston [cut from the bore of a cannon (*2nd ed.*)], were mixed with 50 pounds of black hematite,[7] and were first exposed in

---

[7] Cronstedt, *Mineralogy*, paragraph 204.

a closed crucible for 25 minutes to a fire that was driven by a very large bellows. When the vessel had been opened, it was discovered that a regulus of 201½ pounds had been collected (three-quarters covered by a blackish slag) which was capable of being extended under the hammer to a circular plate ½ inch in diameter before splits appeared at the edge.

In all the following tests the number of the minutes denotes the total time during which the crucible was exposed to the bellows-driven fire.

Expt. 91. When 200 pounds of the same Hällefors cast iron mixed with 50 pounds of calx precipitated from martial vitriol[p] and ignited in a crucible (which made it black and magnetic) had been similarly treated for 20 minutes, it produced a regulus of 206 pounds, embedded in less slag than in the preceding experiment, which was of such ductility that it was capable of being extended into a little sheet having a diameter of ¾ inch without any fissures at the edge. [20]

Expt. 92. Two hundred pounds of Leufstad cast iron endowed with a moderate amount of phlogiston, mixed with 50 pounds of the black hematite previously used, when it had been exposed to the fire for 15 minutes, yielded, in addition to a 168-pound regulus, a multitude of tiny globules all over the sides and even over the lid, disseminated no doubt from a previous overturning of the crucible.[q] The regulus yielded little to the hammer, but all of the globules possessed extraordinary ductility, both the brilliant ones and those on top that were coated in black. The greenish semi-transparent slag was marked with red spots.

[p] Martial vitriol *was ferrous sulphate, and the precipitated calx would initially be ferrous hydroxide, though in air it would turn to ferric hydroxide; the specified heating to turn it black would convert it into the magnetic oxide, Fe_3O_4. The reactions are described in A. F. Cronstedt, Försök till Mineralogie (Stockholm, 1758); English translation (London, 1770), pp. 196–97.*

[q] *To his French translation of this section Grignon appends a note rightly remarking that it was not an accidental inversion but effervescence in the charge that deposited these droplets on the walls and lid of the crucible.*

Expt. 93. Two hundred pounds of the same Leufstad cast iron mixed with 50 pounds of the same martial calx that was used in Expt. 91, after being similarly treated for 15 minutes, afforded a regulus of 222 pounds possessing a ductility approaching that related in Expt. 91. The slag was semitransparent and greenish.

The preceding four experiments were performed in the order in which they are numbered and, indeed, as one crucible was taken out another was at once committed to the hearth. We think this fact should be noted in order that the intensity and increment of the fire in the hearth might be judged.

Expt. 94. Two hundred pounds of the same Leufstad cast iron with 50 pounds of nonmagnetic crocus, prepared from cast iron by calcination after it had been similarly treated for 15 minutes, gave a 210-pound regulus that under the hammer acquired a diameter of $\frac{3}{4}$ inch with the edge still intact. When a drop of nitric acid was deposited on the filed surface and washed off after some minutes, a white spot appeared. It has been noted that the surface of steel covered by nitric acid is blackened, but iron becomes white.[8] The same thing was effected in the manner just mentioned, with a single drop, and it supplies a rapid method for distinguishing between the different varieties of iron. [21]

Expt. 95. Two hundred pounds of the same Leufstad cast iron with 50 of crocus powder prepared at Dylta, after having been similarly treated for 20 minutes, afforded a regulus of 196 pounds that yielded to the hammer. The slag was of metallic brilliance and greenish. It could be filed. It was darkened by nitric acid; in boiling vitriolic acid it gave a black powder that remained intact.

Expt. 96. Four hundred pounds of Dylta crocus in the crucible, mixed with a layer of charcoal dust and with a little bit of borax, when exposed for 20 minutes afforded 68 pounds of pyrites [*i.e., a matte*] of hepatic color, from which it appeared that this crocus

---

[8] The celebrated Rinman in the *Stockholm Proceedings*, 1774. [K. Vetenskaps Akademiens Handlingar, *Vol. 35*, pp. 1–14. An English translation by Alexander Chisholm is given as Appendix A in C. S. Smith, A History of Metallography (*Chicago, 1960*).]

had been born from sulphur. The slag was dark, ashy, and greenish.

Expt. 97. In order that there might be a place for exact comparison, we melted the same Leufstad iron without any addition. Two hundred pounds, when exposed to the fire driven by the bellows in a closed crucible for 20 minutes, produced a black regulus of 196 pounds, yielding little to the hammer and giving a fracture that showed small bright ashen scales. It could be filed, though with a certain difficulty, and a drop of aqua fortis produced a dark spot on the filed surface. When it had been properly heated, it yielded admirably under the hammer and, when hardened in water, gave the fracture of the best steel. No slag.

Boiling vitriolic acid [dissolved the regulus, but (*2nd ed.*)] left a black powder.

Expt. 98. Two hundred pounds of the same cast iron in a crucible whose lid had not been luted, treated in a similar fashion for 20 minutes, gave a regulus of 194 pounds, like the preceding one in all respects. When a little bit of it was boiled in vitriolic acid, it left a small amount of black powder.

Expt. 99. Two hundred pounds of the same cast iron exposed to the fire for 15 minutes in an open but shielded vessel, gave a regulus of 198 pounds, which was more yielding to the hammer than the preceding two, though its fracture was dark and as if full of soot. It could be filed, but it showed a white surface, [22] obscured by very small dark points. Nitric acid made a dark spot. When it was dissolved in vitriolic acid, a smooth dark spongy body remained behind, having the same shape and volume as the little piece that had been put in.

Expt. 100. In order to explore what effect was produced by calcined iron on cold-short iron, we exposed to the fire 200 pounds of the Smolandian iron previously mentioned (Experiment 31), mixed with the 50 pounds of calx used in Expt. 91, for 15 minutes in a closed crucible. The regulus was properly fused and delicately crystallized on the surface. It weighed 180 pounds and was slightly malleable; it showed a white fracture consisting in the little middle part of angular grains but with very fine lamellae toward the outside. It could be filed, and it was whitened by nitric acid. It was ductile when at a red heat; after having been quenched in water it acquired the fracture of steel, though it could be filed.

Expt. 101. Two hundred pounds of the same iron, with a hundredweight of the same ferruginous calx, gave a regulus of 155 pounds, which yielded slightly to the hammer when cold, though it was broken at a red heat. It was not hardened by quenching in water.

Expt. 102. Two hundred pounds of Husaby cast iron, with 50 of iron calx from the same wrought iron from Braås [2nd ed.] heated in a closed crucible for 20 minutes, gave a well-melted regulus and, in addition, a great multitude of little globules on the sides and cover. The regulus broke under the hammer both cold and when at a white heat, but at a red heat it yielded a little. The globules had some ductility when cold, which is also true of those that were obtained in the following two tests.

Expt. 103. Two hundred pounds of the same cast iron, with 50 pounds of calcined Norrberg iron, gave a regulus almost equal to a hundredweight together with scattered globules, all as in Expt. 102. The regulus was cold short, but at a red heat it was more ductile than [23] the preceding and had been richly endowed with a steel-like nature. The globules when melted by themselves without any addition showed no signs of volatilization but had the brilliance of cold-short iron. [A hundredweight of the globules collected in the preceding two experiments, when treated with muriatic acid, gave 49 cubic inches of inflammable air in 15 minutes (2nd ed.).]

Expt. 104. Two hundred pounds of the same cast iron, with 50 pounds of calcined Österby iron behaved in the same manner, and not only was the crucible covered in its lower part with a dark color, but even, as in the two preceding experiments, the inverted crucible serving as a lid was covered inside with the same color [and globules (*2nd ed.*)]. The regulus remaining in the bottom was dissolved by acid of salt much more slowly than the globules and from a hundredweight produced only 43 cubic inches of inflammable air [*2nd ed.*]. When a portion of the same was melted with calx of the same kind of iron, the whole amount remained in the bottom.

Expt. 105. Two hundred pounds of the same cast iron with no addition, melted as the preceding samples, gave a regulus of 197 pounds and showed no signs of volatilization.

Before we proceed further we think that the question raised above must be put to the test, namely whether ductile iron can again become brittle and, if so, in what manner?

Expt. 106. A regulus of $201\frac{1}{2}$ pounds explained in Expt. 90 with the addition of 50 pounds of plumbago was exposed to the fire for 20 minutes in a closed crucible. When the vessel [*vas*] had been broken, the regulus was found to be ashen in appearance, darkened with spotty cavities, and the surface covered with faint and vertical striations. It weighed only 190 pounds. It was hard and was quickly broken under the hammer. The fracture was a whitish ashen color and showed a crystallization similar to the surface. It could be filed. A drop of nitric acid left a dark spot. When a small amount had been dissolved in boiling vitriolic acid, a small amount of black powder remained. It could be forged with difficulty when red hot, and after quenching in water it gave a steel-like fracture.

Expt. 107. The regulus of 206 pounds described in Expt. 91, when packed in charcoal dust and then treated as in the preceding experiment for 18 minutes, produced a globular regulus of 204 pounds[9] with a surface as if burned. It was shattered under the

---

[9] This experiment was later repeated with an appropriate degree of fire and afforded reguli, the weight of which was found to have increased by 1 pound, and sometimes by 2 pounds, in a hundredweight. Hence it is necessary that [*the experiments*] be tempered by a fire of the same duration, in order that nothing might be removed because of calcination. [*Footnote in 2nd ed.*]

hammer, [24] and showed the same behavior as in the preceding experiment with regard to fracture, hardness, and appearance, with both nitric acid and vitriolic acid. When heated red hot, it yielded with difficulty to the hammer: it hardened in water, acquiring a steel-like texture and nature.

Expt. 108. The regulus of 222 pounds described in Expt. 93, when exposed to the fire for 15 minutes in a closed vessel without any addition, was found to have resisted fusion. The color was darkened and it was roughened by certain blisters, but nevertheless, it was as ductile as before. When a drop of nitric acid was allowed to dry out and then washed off, a white spot appeared, as in all wrought iron. A small amount was dissolved in vitriolic acid almost without residuum.

Expt. 109. Seventy-two pounds of the preceding regulus were heated in a violent fire in the same manner for a half hour, but a little piece that had resisted fusion retained its ductility.

Now let us see what changes are produced in cast iron by other materials, of which calcined lead leads the parade.

Expt. 110. Two hundred pounds of Leufstad cast iron, with 142 pounds of lead calx vitrified and pulverized very shortly before lest any acid be present in the air, were exposed to the fire for 18 minutes in a closed crucible. When the little vessel had been broken, a duplex regulus was discovered, leaden at the bottom, to the top of which an additional ferruginous regulus had been joined so tightly that the weight of neither part could be accurately determined. There was a small amount of black slag. The iron was broken under the hammer, giving an obscurely ashen fracture, full of very small scales. It could be filed. A drop of nitric acid produced a dark spot. A small amount cooked in vitriolic acid left an insoluble black powder, which also remained after similar treatment in muriatic acid.

Expt. 111. Two hundred pounds of the same cast iron were similarly treated with 350 of lead calx very recently melted and pulverized, but the effect as to the nature of the regulus was altogether the same. In order [25] that we might know whether the cause of brittleness lay hidden in a small amount of lead contaminating the iron, some of it was dissolved in pure muriatic acid, but when vitriolic acid was poured in drop by drop, nothing was precipitated.[r] The reduced lead had the weight 262 pounds. Some litharge was found sublimed onto the cover.

When they were heated, both reguli (Experiments 110 and 111) broke under the hammer; nevertheless a steel-like fracture and greater hardness appeared after quenching in water.

Let us now consider the effectiveness of calcined manganese on cast iron. It has been noted that black calx of manganese violently craves phlogiston.

Expt. 112. Two hundred pounds of Leufstad cast iron, with 50 pounds of black manganese, exposed to fire for 20 minutes in a closed crucible, gave a regulus of 190 pounds, and it was brittle. A drop of nitric acid left a dark spot. It yielded admirably to a file. Vitriolic acid left behind an intact black powder. The slag was a dark tawny color and full of holes.

Expt. 113. Two hundred pounds of the same cast iron, with 100 pounds of black manganese, when it had been treated like the others for 15 minutes in another trial, gave a perfectly fused regulus of 185 pounds, which indeed yielded a little to the hammer but nevertheless not so much that it could be beaten out into a thin sheet. The fractures were ashen and full of very small scales. The slag was nearly transparent, greenish, and spongy. The regulus could be filed. Nitric acid made a dark spot. When hot, it showed exceptional ductility, and when hardened in water it acquired a steel-like fracture and nature.

Expt. 114. Two hundred pounds of cold-short iron from Braås with a hundredweight of black manganese, exposed in a closed crucible for 15 minutes, gave a regulus of 182 pounds, very fragile and of a very grainy texture, as before, but [26] it was malleable

---

[r] *Had lead been present in the sample, insoluble lead sulfate would have been precipitated.*

both when heated red hot and afterward when it had been cooled. It did not become hard [*on quenching*] in water.

We learned how glass acts from the following test.

Expt. 115. Two hundred pounds of Leufstad cast iron with 100 pounds of crystalline glass,[s] exposed for 15 minutes to the fire in a sealed vessel, produced a regulus of 198 pounds having the entire surface crystallized in parallel striae which were intersected generally at right angles by other parallels. This was broken by the first blow of the hammer. The fracture was ashen in color and full of very small scales. It was bitten by a file, blackened by nitric acid, and dissolved almost without residue by vitriolic acid. It was malleable when hot. It was made very hard by quenching in water, and it showed a fracture equal in nature to the best English steel.

The part of the slag in contact with the melted regulus was darkened and compact, while the outermost part was yellow and foamy.

Calcareous rocks are not infrequently added as an aid to smelting. What kind of effect they have in the crucible we will now put to the test.

Expt. 116. Two hundred pounds of Leufstad cast iron with 67 pounds of calcined chalk [creta usta, *i.e.*, *lime*], exposed to the fire for 25 minutes in a closed crucible, gave a well-fused regulus of 191 pounds which was hard and fragile. Its fracture was an ashen white, full of very small scales. The vitreous slag was a greenish dark color. Heated to a proper red heat, it was malleable to some degree, but at a glowing white heat it broke under the hammer. It possessed a steel-like nature.

Expt. 117. Two hundred pounds of the same cast iron, with $33\frac{1}{2}$ pounds of calcined chalk and the same amount of powdered quartz, exposed in a closed crucible for 20 minutes, produced a

---

[s] *This glass — crystal clear, of course, not crystalline in structure — produced its chemical effect because of its high content of lead oxide.*

regulus of 192 pounds which was a little more malleable than the preceding one. It was of a steel-like nature, for it showed the criteria noted under Expt. 115, and which it is superfluous to repeat here and elsewhere. [27]

Expt. 118. Experiment 116 was repeated in a crucible, the lid of which was not luted. After 20 minutes it gave a regulus of 168 pounds which was perfectly fused but nevertheless possessed some ductility.

Expt. 119. Experiment 117, repeated in a crucible the lid of which was not luted, gave in 15 minutes a mass composed of grains coalesced together but not melted, and so it was exposed to the fire again for half an hour, but in vain. The coalesced grains of metal were easily separated under the hammer, but [*individual grains*] were admirably malleable to any desired degree.

In the last two experiments the intensity of the fire was a little less than in the preceding experiments because, in order that no charcoal dust should fall in and render the conclusion ambiguous, an inverted crucible was put in place to serve as a lid, and there was not a roof of coals as in the tests with the luted lid; to this extent the usual quantity of coals was diminished.

Expt. 120. Two hundred pounds of cold-short wrought iron from Braås, with 25 of burnt lime, exposed to the fire in a closed crucible for 15 minutes, gave a regulus of $191\frac{1}{2}$ pounds, only slightly ductile, with a granular fracture. The slag was black and compact. When heated red hot, it was malleable, but it was not hardened in water.

Expt. 121. Two hundred pounds of the same iron with 100 of burnt lime, similarly treated, gave a regulus of 192 pounds, only slightly ductile. When heated red hot, it yielded a little more. The slag was black.

Expt. 122. Two hundred pounds of the same iron, with 100 pounds of burnt lime, when similarly treated gave a regulus of 191 pounds. The slag was translucent, and green. Breaking the regulus revealed that it was composed of very slender vertical platelets.

Heated red hot it was malleable, but [28] its hardness was not increased in water, and to this extent its original nature remained.

Expt. 123. Two hundred pounds of the same iron with 25 of crude [*unburnt*] limestone, exposed to the fire for 15 minutes with the lid unluted, yielded a regulus of 164 pounds, which was obedient to the hammer a little bit more than the preceding ones but heated red hot showed a slight fragility. The slag was black.

We have also tested the effect of sulphur on the same cold-short iron.

Expt. 124. We have added sulphur repeatedly to 200 pounds of iron at a glowing heat in the crucible and have reduced the whole mass to fusion. We have then calcined the sulphureted iron from a dark brown to a black color and have reduced it to a metallic state with layers of charcoal dust in a crucible. The regulus was fragile, with a fracture like cast iron. It could be filed with difficulty and was darkened by nitric acid; heated red-hot it yielded little to the hammer, but it was hardened by quenching and acquired a steel-like fracture.

We set up the following operation to explore whether iron that is cold short can be corrected by [*admixture of*] hot-short iron.

Expt. 125. We melted in a crucible a hundredweight of Norrberg iron with a hundredweight of Braås iron with a layer of charcoal dust, aided by a 15-minute fire. The regulus weighed 208 pounds. The top was grayish and showed brilliant crystalline unevenness except for a few detached globules. These globules were malleable, although they had been in contact with the charcoal, but the large regulus was quickly broken. The upper part of the fracture appeared much darker and blacker than the lower part, which was conspicuous with shining grains. It was easily filed and was given a dark color by nitric acid. It was not malleable either at a red heat or at a white heat, resembling cast iron. [29]

Expt. 126. A hundredweight of calx from cold-short iron, when reduced in the crucible with layers of charcoal dust, without any addition, gave a regulus of 60 pounds that was fragile and fileable, with a dark fracture displaying a network of metallic lines. The filed surface was darkened by nitric acid, and the entire nature moreover revealed cast iron, for when heated it could neither be wrought nor be noticeably hardened.

Expt. 127. In order to investigate more particularly the efficacy of sulphur on cold-short iron, we exposed to the fire for 15 minutes 83 pounds of the regulus acquired in Expt. 124 with 21 pounds of calcined ductile iron. The regulus was 53 pounds, full of holes, and quite fragile. The fracture was white and crystalline. It was very submissive to the file and was darkened by nitric acid. When heated red-hot, it was a little more ductile than in Expt. 124, and it hardened in water and provided a steel-like fracture. The globules scattered on the sides were ductile. The slag was a dark color tinged with yellow.

Now we will see what the calx of defective iron does to good iron.

Expt. 128. Two hundred pounds of Leufstad cast iron with 50 of calcined cold-short Braås iron, treated in the usual fashion, gave a regulus of 204 pounds which when cold was a little malleable and when red-hot yielded admirably to the hammer, the whole having a steel-like nature.

Expt. 129. Two hundred pounds of the same cast iron with 50 of calcined hot-short iron yielded a regulus of 208 pounds. Half of this when cold was extended admirably by a hammer into a thin plate. After this, the regulus was heated red-hot and tested with a hammer like the other half, but it produced various cracks before it reached the size of the plate forged when cold. Therefore a few pounds of this defective iron is capable of modifying two hundred-weights of good.

(B) We have tested various methods of cementation. That is to say, we have surrounded the principle varieties of iron with various materials — plumbago, burnt calcareous earth [30] (still, however, effervescing a little with acid), and finally even with black manganese calx — in little glass jars which we have then buried in a crucible filled with powdered chalk, and these in turn in a potter's furnace. The fire lasted for 15 hours, and for the last six everything in the furnace glowed at white heat. We examined the contents of the pots after they had been cooled.[t]

Expt. 130. Leufstad cast iron after having been laid in plumbago enjoyed the same weight as before. The little piece was accidentally lost after weighing, and so further examination was not possible.

Expt. 131. A small piece of the same cast iron in calcareous earth was increased in weight more than a hundredth part and was surrounded by a calcined crust. Scale could be knocked off by striking. These fragments adhered to a magnet. When the nucleus had been heated, it was slightly malleable, and it was hardened when quenched in water.

Expt. 132. The same cast iron in black manganese calx gained barely half a hundredth part of weight. It was encased in a crust of calx, which was magnetic like the preceding one. The core resembled cast iron and was neither ductile nor hardenable in water.

Expt. 133. Leufstad wrought iron did not change its weight when it had been exposed to the fire in plumbago. The outside resembled plumbago, but when struck it was malleable, it could be filed very easily, and the filed surface was made whiter by nitric acid. A very slight crust, almost invisible, was similar to plumbago but could be attracted by the magnet. The core possessed the nature of wrought iron.

Expt. 134. The same wrought iron in calcareous earth was found to have been increased by a hundredth part and to be surrounded by a calcined magnetic crust. The outside of the core resembled wrought iron, but the middle, steel.

Expt. 135. The same wrought iron in black manganese calx [31] gained about a hundredth part of weight. A little bit struck with a

---

[t] *Grignon, knowing of Réaumur's experiments on devitrification, appends a note saying that the glass bottles would have been converted to a vitreous porcelain, at least on the outer parts that were in contact with chalk.*

hammer lost its fragile magnetic crust. When carefully filed, it took on a metallic, though dark, luster. The core could be filed, was whitened by nitric acid, and in truth was good ductile iron.

Expt. 136. Österby steel in plumbago kept its weight, surrounded by a little skin similar to plumbago but magnetic. The core possessed the nature of steel.

Expt. 137. The same steel in calcareous earth was increased [*in weight*] by a hundredth part. The outside crust was a red color. The core did not lose any of its steel-like nature.

We also added a little manganese calx, but by an accident we lost this in the furnace.

Expt. 138. Hot-short Norrberg wrought iron in plumbago acquired almost half of a hundredth part [*in weight*]. A little skin like that in Expt. 133 was present. The nucleus was admirably malleable both when cold and when heated. The outermost part had turned steel-like.

Expt. 139. The same wrought iron in calcareous earth gained a hundredth part. The crust was similar to that which was mentioned in Expt. 134. The core, whether hot or cold, was very ductile.

Expt. 140. The same in black calx of manganese gained a further hundredth part and was surrounded by a calcined magnetic crust. The inside was admirably extensible with a hammer, both cold and when hot.

Expt. 141. Cold-short wrought iron from Gränjen, in plumbago, was found to have increased by barely a quarter of a hundredth part and was found to have been clothed with a thin magnetic skin, the outside resembling plumbago. The core did not change its nature — ductile when heated, but cold short.

Expt. 142. The same, in calcareous earth, gained a barely perceptible weight. The nature of the core was found not to have been changed. [32]

Expt. 143. The same, in black calx of manganese, gained half of a hundredth part. The original nature of the nucleus remained unchanged.

Section 6
Corollaries of the Preceding Dry Experiments

Many conclusions arise almost spontaneously from the experiments described in the preceding sections, and we have collected many of them in the following:

(A) Good cast iron, endowed with a moderate amount of phlogiston, is changed into steel by melting alone, both in a closed crucible (Experiment 97) and in an open one (Experiment 98). The same result is reached by adding calcareous earth (Experiments 116, 117), black calx of manganese (Experiment 113), or even crystalline glass (Experiment 115). The addition of calcined lead produced iron (Experiments 110, 111) of a nature at the dividing line between cast iron and steel (Experiment 64). The notable fragility perhaps was derived from the lead, although we have heretofore been unable to detect a trace [*of lead in the iron*]. It is the common experience of workmen that a small amount of lead in an iron hearth may cause brittleness. When steel has been produced, it is found to be not of the same perfection in all cases, but steel born in glass takes the prize from all the others, for it most nearly emulates that steel from England which is commonly called cast steel in regard to its fracture, uniformity, abundance of phlogiston, and other things (Experiment 63).

Good cast iron properly melted with iron calx is malleable and has conferred on it the nature of wrought iron (Experiments 90–94), but with the calx of hot-short iron it retains the same fault (Experiment 129) and with the calx of cold-short iron, produces exceptional steel (Experiment 128).[u]

[u] [*Footnote in 2nd ed. only.*] In assaying iron-bearing minerals by the dry method, it has hitherto been possible to ascertain only the quantity of the metal, but in order that the quality also may be distinguishable (without doubt a thing of great importance), the strength of the ferruginous calx already noted (Expts. 90–94) allows this end to be easily obtained. Consider the method that is used in the Laboratory of Uppsala. The regulus is generally obtained with nothing added in the crucible except a layer of charcoal dust,

Husaby cast iron, which in the forge gives rise to cold-short iron, [33] has shown some peculiar phenomena of volatilization, but although this was not corrected with calx of cold-short iron (Experiment 102), with calx of hot-short iron it acquired a steel-like nature (Experiment 103), and it is hardened in water like a regulus made with calx of good iron but is very fragile (Experiment 104).

Cast iron is hardly changed by cementing in plumbago (Experiment 130) or in black calx of manganese (Experiment 132), but in powdered limestone[v] it attains the nature of wrought iron (Experiment 131).

(B) Ductile wrought iron melted with charcoal dust (Experiment 107) or with plumbago (Experiment 106) acquired steel-like properties.

Nothing was effected by cementing [*ductile wrought iron*] in plumbago (Experiment 133) nor in black calx of manganese (Experiment 135), but in powdered limestone the inmost part acquired the nature of steel (Experiment 134).

---

or at most only an equal weight of fluorspar and a half weight of clay (cf. *Opuscula*, Vol. II, pp. 195–96). The regulus obtained is always broken under the hammer, but when it has been crushed to powder and properly melted with a quarter of ferruginous calx, it produces its true nature. If, after this, it is shattered [*on cold hammering*], it contains cold-short iron, but if it is malleable the defect of hot shortness may nevertheless be present, as may be determined in the following manner. The regulus may be flattened out a little when cold, while it is held by a small tongs, and half of it carefully extended by forging until the edge begins to make cracks. When this has been done, the hammered-out part is held by the tongs, and the other half may be embedded in hot coals and, when properly red, tested with a hammer. If it can be extended uniformly and as much as when it was cold, then it is good iron; but if less well than before, it has been corrupted by the hot-short quality, proportionately more as the edge is more quickly split.

[v] *Bergman uses* pulvis calcareus *and* terra calcarea *as equivalent, though the former seems to imply an artificially and the latter a naturally comminuted material. The experiments do not show what would be expected today, for the black calx of manganese* (MnO_2) *should have been a far more effective oxidant than limestone. That powdered limestone should have converted iron into steel while plumbago in the next paragraph did not is equally inexplicable.*

(C) The nature of cold-short wrought iron is changed little. Melted with iron calx and then heated red-hot, it became malleable and was slightly so even when it was cold (Experiment 100), but with a larger amount of calx it seemed to gain the nature of hot-short iron (Experiment 101). When melted with black calx of manganese, it is corrected a little bit (Experiment 114), but with powdered limestone its original cold-short quality remains (Experiments 120–123). Cold-short iron that had been made by sulphurization, calcination, and reduction acquired a steel-like nature when melted with calx of good iron (Experiment 127).

Nothing is effected by cementing [cold-short iron] in plumbago (Experiment 141), black calx of manganese (Experiment 143), or powdered limestone (Experiment 142).

(D) Hot-short wrought iron was not corrected by cold-short iron, or vice versa (Experiment 125).

The fault of this iron seems to be taken away by cementing in plumbago (Experiment 138), in powdered limestone (Experiment 139), and in black calx of manganese (Experiment 140), [34] a thing that is most worthy of note, since the calx of this iron corrupts good iron with its fragility when melted with it and heated red-hot (Experiment 129).

(E) Steel keeps its own nature when cemented in plumbago (Experiment 136) and also in powdered limestone (Experiment 137).

(F) Now let us investigate the reasons for these phenomena as far as it can be done. We know in advance that the amount of phlogiston that accompanies the various mutations of iron according to the limits previously discovered is either increased or is lessened under the present circumstances. Thus cast iron, which produces $39\frac{1}{2}$ cubic inches of inflammable air from a hundredweight (Experiment 2), if it is made ductile with iron calx in the crucible, then gives 51 cubic inches (Experiment 4), but if again melted with the assistance of charcoal and imbued with a steel-like nature, produces 45 (Experiment 62). Similarly, cast iron melted with calcined lead produces 43 cubic inches (Experiment 64) and with glass, 45 (Experiment 63).

As to that pertaining to the combined caloric matter, its quantity varies to such a degree for each of the different states (Section 7, C) that under present circumstances the iron releases as much or even more than it had fixed in the previous operations.

Moreover, from Section 9 we have recognized certain principles as present in any of the states of iron, though with differences in the quantity of them. That is to say, ductile iron contains almost no plumbago, but it enjoys a great abundance of both combined caloric matter and the inflammable; steel enjoys a lesser amount of the latter, but it is richer in the former, and cast iron saturated with plumbago possesses the smallest amount of phlogiston and a small amount of caloric matter.[w]

Therefore in order that cast iron may become ductile, plumbago must be driven away or decomposed, and at the same time the remaining very subtle principles [the inflammable principle (*2nd ed.*)] must be increased. For the smelting of iron in the hearth is completed with a very intense fire and a very violent blast, by which the principles of plumbago, which is composed of aerial acid and phlogiston [35] (Section 8, D) are driven away; first it [*the phlogiston*] flies away, and later, unless we err, it becomes affixed to the iron molecules. [From this it is understood why iron saturated with phlogiston remains liquid longer and congeals reluctantly. That is, the abundance of phlogiston impedes the decomposition of the plumbago, and as long as this is present the mass is more fusible, but if this has been removed, the same degree of fire no longer suffices for fusion, whereupon even iron soon begins to solidify. A disturbance like boiling and a shining incandescence

---

[w] *In the second edition this sentence reads as follows:* "That is to say, ductile iron contains almost no plumbago, but it enjoys a small amount of combined caloric matter and a great abundance of the inflammable; steel has a lesser amount of phlogiston but is richer in caloric matter; and cast iron, saturated with plumbago, possesses little phlogiston and a great amount of caloric matter."

proclaim this destruction [*of plumbago*], the former arising from the departure of the aerial acid, the latter from the copious phlogiston seizing the vital air and generating light.[x] The addition of slag aids the operation a great deal through the mediation of the calcined iron with which the slag is more or less burdened.][y] The question may be put: Why does cast iron, even when copiously surrounded by charcoal, receive only so much phlogiston as is sufficient to confer ductility? We answer that plumbago, remaining always an acid combined with phlogiston to saturation, is a species of sulphur. But it is necessary that metals release a portion of phlogiston before they can be joined with common sulphur so that no marriage to other things can take place. Therefore isn't it permissible to make the same conjecture about plumbago? And certainly experience directly proves that, for up to now we have found no cast iron without plumbago, and none in which the iron molecules per se enjoy the abundance of phlogiston that corresponds to that in the ductile iron that is prepared from the cast iron and which even varies as the state of the iron is changed. Moreover the amount of the inflammable principle in the different types of both cast and ductile iron seems not to have been confined within narrow limits but wanders about with some latitude (Section 4, C–G), regardless of whether this variation depends, as indeed seems most likely, upon the circumstances of the process or upon certain discrepancies in the nature of the iron molecules themselves.

It is indeed beyond doubt that plumbago when present in iron is capable of being decomposed by exposure to air by a suitable method, but nevertheless it does not follow therefrom how it may turn out that cast iron may be spoiled by melting even in a closed crucible with additions of iron calx alone. It is necessary to note

---

[x] *Note that Bergman fully understands the refining of cast iron in a strong blast of air and comments on both the evolution of bubbles of aerial acid (CO_2) and the brilliant incandescence accompanying the reaction.*

[y] *The material in brackets is present in the second edition only.*

here that iron calx is capable of absorbing aerial acid, and for our experiments to succeed, it must surround the particles of cast iron powder to such an extent that it may be present both at a white heat and at the moment of fusion, seeking the aerial part of the plumbago, while at the same time the iron molecules try to seize and saturate themselves with the phlogiston united to it. Plumbago is decomposed by this twofold attraction, the iron is enriched by the inflammable, and if anything is left over, the portion of iron calx [36] is reduced proportionately. In order that this may be apparent to the eye, let us present the most difficult of our experiments. Two hundredweight of Leufstad cast iron with half this weight of iron calx gave a ductile regulus of 222 pounds (Experiment 93). Two hundredweight of this cast iron carries 6.6 pounds of plumbago (Section 8, D, Experiment 342 [*sic., actually Experiment 210*]). An amount of this plumbago that fully decomposes about 5 pounds of niter also will turn to alkali barely half a pound of niter.[z] Therefore one part of this plumbago contains as much phlogiston as 10 parts of iron, and 6.6 of plumbago as much as 66 of iron. Moreover, it was brought out in the preceding experiments that a hundredweight of Leufstad cast iron produced 39.5 cubic inches of inflammable air (Section 3, Experiment 2) and the regulus made ductile in the crucible afforded 51 (Section 3, Experiment 4), which shows a difference of 11.5 inches between cast and ductile iron, or 23 cubic inches for our case of two hundredweight. Now let us see if the phlogiston of 6.6 pounds of plumbago may suffice to reduce 22 of iron calx and thus supply the

[z] *This is an ingenious attempt to measure the relative amounts of phlogiston, although numerically the results are quite different from what would be expected from the reactions involved. The equation $5C + 4KNO_3 \rightarrow 5CO_2 + 2K_2O + 2N_2$ corresponds to one pound of carbon reacting with 6.73 pounds of niter. Similarly, $15Fe + 8KNO_3 \rightarrow 5Fe_3O_4 + 4K_2O + 4N_2$ has one pound of iron combining with 0.96 pounds of niter. A pound of carbon therefore possesses 7.4 times the capacity for combining with oxygen; i.e., it has 7.4 times as much "phlogiston" as does iron — a ratio not far from that found by Bergman. The text is not clear: perhaps 6.6 pounds of plumbago decomposed 5 pounds of niter, but if this is so the quantitative amounts are even further from what we would expect.*

gain [*in weight*] that was mentioned above. In the manner we have indicated, 66 pounds of iron calx is capable of being reduced by 6.6 pounds of plumbago, and so 22 is subtracted from 66 and 44 remains. We have previously demonstrated that as much phlogiston is present in one cubic inch of inflammable air as is contained in about 2 pounds of wrought iron (Section 4, B). Therefore 44 pounds of iron in which an abundance of phlogiston is present corresponds to 22 inches of inflammable air [*compared with 23, the difference found experimentally between wrought and cast iron*]. One cubic inch is still lacking, but it may be observed that one part of iron in order to be calcined requires a little less than half a part of niter [*and if it has been forged by the common method, it gives only 50 cubic inches (Experiment 3) (2nd ed.)*]. We believe that a more valid confirmation of the explanation could hardly be expected.

The fact that such small portions of iron calx were reduced in Experiments 90, 91, and 92 must undoubtedly be derived from the surroundings through which more or less inflammable matter was capable of being released without effect. In assaying [37] iron ores even in crucibles in which the joints with the lids have been most skillfully sealed, it is very rare that all of the metal is reduced even though phlogiston abounds.

We will discuss defective iron — both hot short and cold short — in another place where we will consider the experiments cited a little while ago, as well as more, to be described, that pertain to both types; here we consider good iron only.

Now that the extreme states of iron have been considered, the intermediate state of steel causes less difficulty. Once the analysis of the steel has been understood, it is allowable for the processes by which the various kinds are prepared to differ greatly, though they ought to agree in regard to the fact that the particles are closely joined together and are provided with a proper amount of the inflammable principle and of plumbago (Section 9). The desired effects are clearly apparent to the eye, while the operation [*of steelmaking*] is accomplished by repeated smelting, heating, and forging, for smaller masses are then treated by a more intense fire and bellows than when it is intended to make the iron ductile and expose it to more air.[aa] It seems that an effective change is less to be expected in a closed crucible; nevertheless, steel is also generated here, even when melting cast iron without any addition (Experiment 97) or with glass alone (Experiment 115). If therefore a simple fusion destroys a very great part of the plumbago, we will not wonder that the same result occurs with additions of limestone (Experiments 116, 117), or powdered calcined manganese (Experiment 113). But even if the first observation has been solved, a very difficult perplexity simultaneously arises. For ductile iron with charcoal dust goes into steel (Experiment 107) and with plumbago into cast iron, a neighbor of and related to steel, which seems to add weight to the opinion held by many concerning the larger amount of phlogiston in steel than in iron. However, by analysis, there is found to be very little directly inflammable matter in steel (Section 4, E), while almost no plumbago occurs in ductile iron, though in steel the insoluble residuum is notable enough (Section 8, D). [38] Let us further consider whether, with the phlogiston that has been added in these cases, aerial acid is simultaneously present which on melting enters the fabric of the iron and seizes a portion of the removable phlogiston in order that the plumbago that this gives rise to may be able to mineralize the iron [and increase the absolute weight of the same by one or two parts in a hundred — compare the footnote to Experiment 107 with Section 9, C (*in 2nd ed. only*)]. Steel can be produced both from cast iron and from wrought iron. In the former case a portion of the plumbago must be destroyed in order that the iron molecules may be properly

---

[aa] *In the second edition this sentence reads:* "The desired effects are clearly apparent to the eye, while cast iron is repeatedly worked by smelting, heating, and forging, for small masses are treated by a very intense fire and bellows, almost as if the iron was to be made malleable, and they are exposed to a large amount of air."

enriched with phlogiston, but in the latter case the superfluity must be joined to the aerial acid in order that the same molecules may become poorer by a sufficient amount. The limits are so definite that hardened steel, and the same softened by the fire, produce the same profusion of inflammable air (Experiments 12, 13).

The degree of cementation that is attained undoubtedly differs with respect to the abundance or lack of phlogiston in the cement, but meanwhile we learn from the experiments that some amount of inflammable air is always dissipated, for small pieces subjected to these tests have been encased with crusts of calcine, both in plumbago (Experiment 133) as well as (though far more slightly) in limestone powder and in black manganese (Experiments 134, 135). Hence we have concluded rightly that charcoal dust surrounding ductile iron in a closed vessel and exposed for a long time to a degree of fire almost sufficient to bring about fusion does indeed produce some amount of aerial acid. With the superfluous phlogiston this aerial acid generates plumbago which is trapped in the fabric [*of the iron*] and imperfectly prevents the drying out or dephlogistication of the interior, whence not infrequently, when the steel is dissolved, many molecules that have been burned remain unattacked by vitriolic acid but require muriatic acid. Aren't these [*burned*] things separated when the fusion is being carried out, leaving the nature of the steel unchanged? And isn't the very great perfection of English [*steel*] associated with this deprivation? It has also been demonstrated previously that ductile iron necessarily loses a portion of phlogiston on being changed into steel.

The fact that wrought iron keeps its ductility in plumbago [39] (Experiment 133) but in limestone dust acquires in its inner part the nature of steel (Experiment 134) signifies well enough that there is no need for anything inflammable outside, but the superfluous amount that is present in ductile iron enters a plumbaginous union with the aerial acid, where this acid is not yet bound up by the cement. Indeed by fusing it with plumbago even cast iron becomes close to steel in its nature (Experiment 106), which hardly happens in cementation unless perchance it may be necessary to heat for a longer time than was used in our experiments. The steel-like nature that was present only in the inner part [(Experiment 134) (*2nd ed.*)] admits of a twofold explanation, for either the change begins within, progressing to the outside, and had not yet reached the outer parts, or else, as in truth is more likely, in the beginning the entire sample has been made steel-like and then, when the calcined crust was formed, recovered its original ductility on the edge [which is in perfect accord with the tests of Réaumur (*2nd ed.*)].

For the rest, our experiments regarding cementation undoubtedly would have given more light if we had been permitted to vary the duration of heating and to use a better-sealed vessel [*2nd ed.*].

Section 7
The Quantity of Caloric Matter in Iron

It can scarcely be doubted that heat and its phenomena arise from a peculiar matter — at least it is not easy to explain on the basis of an internal movement of the particles constituting bodies how a great fire could often have its origin in a tiny spark. As daily experience testifies, the communication of movement proceeds with diminishing [*force*], but fire left to itself increases as it propagates, provided that it does not lack suitable nutriment. Without a doubt an immense difference of movement is found between a first spark and a burning house or city, and the effect surpasses the cause by an astonishing amount. We omit other arguments since a complete demonstration does not belong here. [40]

We also think that it is in conformity with the truth that in any body whatsoever there may be present more or less caloric matter that has been bound by the force of attraction, that it lacks the power of making things hot as long as it remains bound, but which nevertheless is quickly recovered once it is freed. This is much like an acid whose principal properties lie hidden in perfect neutral salts, though it exercises them again when it has been freed.

Notable differences in respect to the quantity of this caloric matter change the states of many substances to such a degree that a thing that is now solid may at another time become liquid, indeed it even may be able to acquire fluidity.[10] The phenomena testify that a definite amount [*of heat*] is necessary to whatever state [*a substance may be in*] whether in regard to the mutual mobility of its particles or to some different quality of them. This amount may be called specific heat, for in truth it accompanies the nature of all bodies.[bb] This work, this labor, was done in order to determine it accurately. Indeed, we know that extraordinary tests have been performed in our times, but we frankly confess that the methods proposed seem to us not only frequently to be inaccurate but even (unless we deceive ourselves) sometimes false, a matter with which we will, perchance, concern ourselves on another occasion. [*Second edition reads:* We frankly confess that the methods avail-

---

[10] The older writers observed the distinction, which is neglected by more recent ones, between the names "liquid" and "fluid," the latter not encompassing vapors and smoke.

[bb] *In the important section that follows, Bergman shows that he had a good feeling for the principles of thermochemistry, for he saw that there was energy associated with matter in a specific way and that its evolution and absorption accompanying changes of state and chemical reaction could be quantitatively studied. His experimental methods, however, were far too crude to give useful results. Bergman's use of the term "specific heat" is not that of the modern phrase but refers to essentially the (negative) heat of formation, referred to the uncondensed elements at room temperature as a standard state. Bergman had read a paper on latent heat in 1772, and in 1781, the same year as the present essay, he published an important treatise on specific heat (K. Vetenskaps Akademiens Handlingar, 1781, 2, 49). It is interesting to note that young Gadolin, who "defended" this paper as part of his doctoral examination, continued to work on thermal effects and published on true specific heats in 1784. (See D. McKie, The Discovery of Specific and Latent Heats [London, 1950].)*

able require so much time that we have not yet been able to determine the specific heat of each variety of iron.]

(A) Meanwhile, while these undertakings are being matured and reduced to certainty, when it is asked in the present case whether iron in the state of cast iron, steel, or wrought iron possesses the same specific heat, we have directly observed with a thermometer the increase in temperature that occurs when a metal is dissolved [*in acid*], both metal and acid having been kept at the same temperature as the solvents for at least a day and a night. We do not, indeed, think that this way is as trustworthy as it is easy, but we have used it for want of a better one [for unequal pulverization, inherent heterogeneities, and other things easily impede the action by presenting less of the stuff to be dissolved in the same area of the solvent. Moreover, the rate of solution is influenced by the different states of the metal and different varieties of the same state. But if the obstacles are removed as much as possible, the method now to be presented can determine (the temperature) in a more or less definite manner (*2nd ed.*)]. Consider an outline of the matter that occurs to us. [41] The binding of the molecules is released by the solvent and new unions are formed which, if they are greater [*i.e., more strongly bound*] than the ones destroyed, fix a [*greater*] quantity of caloric matter; it is necessary, in order to restore the equilibrium, that the heat in the surroundings be diminished, and therefore the mercury in the thermometer subsides; but if the [*new bonds*] are smaller [*i.e., weaker*], the surplus is freed and makes the surroundings hot, whence the liquid in the thermometer rises; finally, if the new bonds demand precisely the same quantity of heat (which rarely happens), no change will be seen in the thermometer. A certain amount of caloric matter does indeed cling to the small molecules of bodies, especially of moving bodies, in the manner recently explained, until a stronger attraction changes this connection. He who thinks that everything is permeable to heat and, with the great Newton, advances the idea that at least half of the volume of even the heaviest of all materials, gold, is occupied by invisible holes and that the magnetic effluvia are disturbed in no manner by the interposition of this very dense metal, will find no contradiction in similar atmospheres surrounding primitive molecules, even in the solid state. These same molecules increase their movements about when they have been separated by a sufficient amount of heat and by their greater contact fix a greater quantity of caloric matter. Thereby they become more mobile and achieve liquidity, and when they have been penetrated by a still greater portion [*of caloric*], they eventually become a wholly vaporous fluid. From the above experiments it therefore appears that the indication of temperature differs from the ambient only by that [*heat*], either bound or liberated by the action of the solvent under present circumstances. [*It is indicated*] more accurately as the solution is completed more swiftly without using external heat, for if the decreases or increases are produced at the pace of a tortoise, they can be easily absorbed or dissipated without any noticeable variation in the thermometer.

It may be permissible to illustrate the matter further by an example, recently advanced and to some extent explained elsewhere.[11] If a powder of any neutral salt is poured into water, [42] we will see that the mercury of a thermometer in the solution goes down, since the new state of liquidity requires more heat than the state of solidity that has been lost, and the very small molecules being separated increase their attractive force [*for the solvent*] because of their far greater surface. It is also apparent that when a great mass of crystals quickly congeals, the mercury in the thermometer rises more or less, for caloric matter has been released; since the area has been decreased, the [*molecules'*] contact and power of attracting caloric matter has been diminished. But there are gradual differences depending on the different natures of salts, and so far the effect is observable only under favorable circumstances.

[Another more definite case occurs when acids dissolve metal:

---

[11] *Opuscula*, Vol. I, pp. 234-36.

the metals must first have been dephlogisticated to some extent, then the calces must be joined to the solvent and finally the new unions must be taken into the water. These changes are all completed at almost the same moment, and each of them varies the specific heat. The quantity of heat in the water usually exceeds that contained in the acids, and similarly the supply in these acids is richer than that of the metals. It follows then that on solution, the water and acids release more heat than is required by a strongly bound metallic salt (*2nd ed.*).]

But we will now devote ourselves to the experiments we have proposed.

(B) We always used equal weights of the solvent and of the thing to be dissolved. These were placed with the apparatus nearby for an entire day. Acid was poured into the little vessel that contained powder or filings of the iron that was to be investigated, the thermometer bulb was immersed, and the rise of the mercury carefully noted. The ambient temperature was being checked at the same time with another thermometer. When aided by no external heat, vitriolic acid acted so slightly upon the iron that the increase in heat amounted to barely one or two degrees; muriatic acid excited it a little more but in no way enough; but nitric acid produced more notable effects, with swelling effervescence. In order that the violence of this foaming may be described in a definite way, we have indicated such violence that ascends to the upper rim of the little vessel by 4, that which ascends about three-quarters of the height by 3, half by 2, and finally that which reaches one-third by 1. The mixture by itself occupies about a quarter [*of the vessel*].

The same nitric acid as in Section 3, D, was used, but only a half measure, since only a half of a hundredweight was devoted to the tests, for a double quantity [43] of iron, producing a double increase in heat, would demand a very large scale. [Moreover, since nitrated iron ought to rise the same amount in all cases, it is evident per se that the specific heats of the things to be examined are inversely proportionate to the degrees of temperature increase, other things being equal. (*2nd ed.*)]

In the following table only the effects of nitric acid have been noted. The first column indicates the ambient temperature [*in degrees Celsius*], the second the maximum degree of temperature rise, the third the difference between the two, and the fourth the violence of the foaming.

[Experiment no.]	[Kind of iron]	Ambient temperature [°C]	Max. temperature after reaction [°C]	Difference [°C]	Degree of foaming
Expt. 144.	Leufstad cast iron, filled with phlogiston	+11	+37	26	2
Expt. 145.	The same Leufstad cast iron	2	28	26	2
Expt. 146.	Leufstad cast iron with a moderate amount of phlogiston	11	21	10	1
Expt. 147.	The same Leufstad cast iron	5	15	10	1
Expt. 148.	Wrought iron from Leufstad	11	72	61	4
Expt. 149.	Wrought iron from Österby	11	79	68	4
Expt. 150.	Steel from Österby	13	63	50	4
Expt. 151.	Forsmark cast iron	14	24	10	1
Expt. 152.	Wrought iron from Forsmark	14	81	67	4
Expt. 153.	Steel from Forsmark	14	60	46	4
Expt. 154.	Brattefors cast iron	14	33	19	2
Expt. 155.	Wrought iron from Brattefors	14	75	61	4
Expt. 156.	English steel, much praised	13	70	57	4
Expt. 157.	Steel made by Dr. Quist	13	65	52	4
Expt. 158.	Hällefors cast iron, saturated with phlogiston	13	39	26	3
Expt. 159.	Hällefors cast iron, poor in phlogiston	13	30	17	2
Expt. 160.	Steel from Dalia iron, well endowed with manganese	14	56	42	4
Expt. 161.	Wrought iron from Norrberg, hot short	14	79	65	4
Expt. 162.	Wrought iron from Gränjen, cold short	14	81	67	4
Expt. 163.	The same iron, 25 assay pounds	14	48	34	4
Expt. 164.	Husaby cast iron	13	34	21	2

[Experiment no.]	[Kind of Iron]	Ambient temperature [°C]	Max. temperature after reaction [°C]	Difference [°C]	Degree of foaming
Expt. 165.	Cold-short wrought iron, made from the preceding cast iron from Braås	15	77	62	4
Expt. 166.	The same melted in a crucible with half its weight of lime (Expt. 122)	12	85	73	4
Expt. 167.	The same iron melted with limestone; malleable	15	77	62	4
Expt. 168.	Steel made from brittle iron from Braås	13	50	37	2
Expt. 169.	Steel originating in glass (Expt. 115)	13	66	53	4
Expt. 170.	Steel arising from smelting alone (Expt. 97)	13	56	43	3
Expt. 171.	*Ethiops martialis*, black, completely magnetic. Purchased at a pharmacy	14	14.5	0.5	0
[45] Expt. 172.	Calcined chalk carefully washed and dried, half a hundredweight	14	20	6	2
Expt. 173.	The same well burned and not slaked, half a hundredweight	14	84	70	

(C) If the experiments described in the preceding section are considered, it appears without difficulty that every state is limited with respect to the heat that it conceals. Thus the lower limit for cast iron dropped to 10 degrees, the highest rose to 26. Among the seven varieties of cast iron that were tested, those that had been deprived of phlogiston approached the bottom, but those filled with inflammable air or those varieties that produced cold-short iron tended to the other extreme.

The six kinds of steel produced greater heat; among them none excited a degree of heat less than 37 and none more than 57.

We have not yet discussed the nine specimens of wrought iron. Although these were more numerous, their variation fell between narrower boundaries than the preceding groups, producing however very great heat — not less than 61 degrees and not exceeding 68. On account of the sluggishness of its dissolution, native

Siberian [*meteoric*] iron could produce almost no increase of heat. Its faulty nature reveals only little of the heat that is concealed in it.

Moreover, other things being equal, we see that a double quantity of iron produces a double increase in temperature (Experiments 162, 163), and when the fusion of the iron was repeated, it did not therefore produce a more intense heat but only the amount adapted to the state of the metal (Experiments 156, 157, 169, 170). Only defective cold-short iron wandered a little bit beyond the limits (Experiment 166). [46]

It is also worthy of note that the same rise in temperature was generated in places enjoying different degrees of ambient heat (Experiments 144–147).

Therefore since in the experiments described only the nature of the iron was varied, it is at once evident that the difference in the effects must be derived from that. The two final experiments, set up with calcareous earth (Experiments 172, 173) have therefore been added for comparison.[12]

Section 8
Foreign Substances That are Present in Iron

We must now consider the peculiarities of the heterogeneous substances which are often associated with our metal, or are supposed to be, and which nevertheless are not pertinent to its essence although they can impart peculiar properties.

Such materials are manganese, arsenic, zinc, plumbago, and acid of sulphur, which therefore may be examined separately.

(A) In order that manganese may be easily distinguished when it is present, we have used the following method, the most expeditious that we have been able to discover and which was recently described elsewhere.[13] In this let a small amount of iron ore or of the metal itself be made properly red-hot in a crucible

---

[12] An exceedingly small portion and a sluggish thermometer gave rise to an error appearing on p. xii in the beginning of the first volume of the *Opuscula*.

[13] *Opuscula*, Vol. II, p. 225. [*Eng. trans., Vol. II, p. 228.*]

exposed to fire, then let there be added five times its weight of purified niter and, when the effervescence has ceased, the crucible is removed. After it has been cooled, if there is manganese present, there is often found inside, extended right up to the rim, a little green skin, or one turning toward green from bluish. If [*the sample*] yields only feeble traces, a second addition of niter equal to the first may be made, by which the smallest amounts are extricated. [47]

Ashes and charcoal dust should be carefully kept away. The rationale of this method of exploring is this. Manganese, when it is calcined and dissolved in molten niter, becomes blue, but when iron is present at the same time this color turns green because of the admixed iron calx[14] [which more or less turns it yellow (*2nd ed.*)].

Here also we have submitted to the test the varieties of iron previously examined, and although only three assay pounds were used of each, the results of this procedure are distinguishable and tangible.

In order to avoid prolix descriptions, we first used Eisenerz iron as a basis of comparison, which iron was explored by a method described elsewhere,[15] and contains in a hundredweight about 30/100 of manganese.

Expt. 174. Eisenerz iron was treated with niter. In the first application the entire inside of the crucible was covered with a little skin of an intense green which, if anything, deepened upon the addition of the second amount.

Expt. 175. Leufstad cast iron, abounding in phlogiston, produced a small amount of green on the first application [*of niter*], but with the second the color came out even more strongly.

Expt. 176. Leufstad cast iron, endowed with a moderate amount of phlogiston, did not become visibly green with the first applica-

[14] *Opuscula*, Vol. II, p. 220. [*Eng. trans.*, Vol. II, p. 223. The *lutescit* of the interpolation in the second edition that we translate "turns it yellow" could also mean "muddies it." Mn^{++} is, like Fe^{+++}, yellow in alkaline melts, but more oxidation would give purple Mn^{+++}, almost black when Fe is also present in quantity.]

[15] *Opuscula*, Vol. II, p. 228. [*Eng. trans.*, Vol. II, p. 232.]

tion [*of niter*] but with the second showed a little skin that was blue turning to green.

Expt. 177. Leufstad wrought iron gave the same behavior as the preceding experiment.

Expt. 178. Åkerby cast iron itself turned a little bit green with the first application [*of niter*], but the sides of the vessel were scarcely colored. With the second application all the inside of the vessel turned blue.

Expt. 179. Åkerby wrought iron turned green in a little while on a single application, together with the sides. [48]

Expt. 180. Ullfors cast iron similarly stained the sides with green on the first application.

Expt. 181. Ullfors wrought iron afforded no distinct evidence of manganese with the first application but with the second behaved in the same manner as the preceding one.

Expt. 182. Österby wrought iron acted in the same way as did Åkerby cast iron.

Expt. 183. Österby steel showed the same phenomena.

Expt. 184. Forsmark cast iron spread out a distinct greenish color with the first application, and this was increased a little by the second.

Expt. 185. Forsmark wrought iron behaved the same way.

Expt. 186. Forsmark steel a little more weakly.

Expt. 187. Brattefors cast iron, as Forsmark.

Expt. 188. Brattefors wrought iron, as Ullfors wrought iron.

Expt. 189. Hällefors cast iron, overloaded with phlogiston, showed no traces of green with the first application [*of niter*] but with the second showed distinct traces.

Expt. 190. Hällefors cast iron, impoverished in phlogiston, behaved similarly.

Expt. 191. Native Siberian iron with the first application brought forth some but with the second, an extraordinary green color.

Expt. 192. Cold-short Norrberg iron did not turn green with the first application but became distinctly so with the second.

Expt. 193. English steel turned green with the first application, the sides only a little, but these nevertheless put on the same color with the second application.

Expt. 194. Steel prepared by Dr. Quist behaved as did the English steel, though it acquired a slightly more intense color.

Expt. 195. Steel made from Dalia iron behaved as the preceding type. [49]

Expt. 196. Husaby cast iron brought out the color with the first application.

Expt. 197. Cold-short wrought iron made from Husaby cast iron turned green only with the second application.

Expt. 198. Steel prepared from the same iron behaved as the preceding type.

Expt. 199. Cold-short wrought iron from Braås showed green with the first application, and this increased a little with the second.

Expt. 200. The same iron, rendered malleable with the help of limestone additions, tinted the sides more distinctly on the first application.

Expt. 201. Manganese dissolved in acid, either vitriolic or nitric, and freed from copper by the action of zinc, produced an intense greenness with the first application.

Expt. 202. Scrapings of copper behaved almost the same.

We have concluded from these tests that each variety used contains manganese, and if the experiments performed by others[16] are considered at the same time, there emerges as a probable conjecture that manganese is never, or at least very rarely, removed completely from iron by the usual operations; that defective ductility, either in an incandescent state or at ordinary atmospheric temperatures, is not to be ascribed to the absence of manganese; and that manganese when present in a very great quantity neither prevents ductility (Experiment 174) nor does it always produce it (Experiments 175, 176, 178, 180, 184, 187, 196).

[16] The celebrated Hjelm in the [*K. Vetenskaps Acad. Handlingar*, 1778, *39*, 82].

If anyone should be pleased to think that these green-producing phenomena are derived from copper, it is necessary that he recognize its presence in every variety examined, but, unless we are mistaken, this is groundless. Without a doubt, two different metals are able to bring out the same color with niter, although they differ in all their other properties. Glass acquires a red tint both from copper and from iron, [50] but these two metals are nevertheless very distinct from each other. To be sure, it is true that manganese is nearly always contaminated with copper, but if this foreign matter is precipitated by the action of zinc, the purified manganese colors the niter the same as before (Experiment 201).

As to the amount of manganese that is contained in iron, we place this at 30/100, indeed generally less than this, since Eisenerz iron has this amount, and it is without doubt far richer than the other varieties (Experiment 174). But in order that the amount of color extracted may be estimated with certainty, an agreement of very many experiments is required, for small variations, easily escaping the attention of the operator, often help or hinder the efficacy of the niter. Thus in the preceding experiments cast iron turned green more weakly than did wrought iron (Experiments 178, 179). The quantity must therefore be determined in another more exact manner. Vinegar (the distilled variety) dissolves black calx of manganese,[17] but it dissolves only a little iron [*2nd ed.*].

Expt. 203. A hundredweight of well-calcined Österby iron was immersed in well-distilled vinegar and for some days exposed to a digesting heat. [*Manganese*] was then precipitated from the solution by fixed aerated alkali in the form of white calx [*manganese carbonate*] which, after it had been dried out, weighed barely one pound, corresponding to about half that weight of reguline manganese.[cc]

---

[17] *Opuscula*, Vol. II, pp. 219, 453. [*Eng. trans., Vol. II, pp. 222, 469.*]

[cc] *Grignon remarks in a note to his translation that the precipitate cannot be manganese alone, for later experiments show that the vinegar solution sometimes gives Prussian blue with "phlogisticated alkali," i.e., potassium ferrocyanide.*

Expt. 204. A hundredweight of calcined hot-short Norrberg iron, after being similarly treated, afforded the same quantity of aerated manganese.

Expt. 205. An equal weight of calcined cold-short Braås iron gave by the same method almost five pounds of martial precipitate which, with niter, turned a little green. Phlogisticated alkali precipitated an abundance of Prussian blue, although the same alkali produced hardly any blue with the preceding two solutions. [This difference is illustrated in the following (Experiments 264–268). (*2nd ed.*)] [51]

(B) It is well known that arsenic when present in iron produces brittleness, and it also resists removal with great pertinacity, though this is not sufficient to prove that the cause of cold shortness is generally derived from it. Therefore we have sought illustration in ores. That splendid ore that occurs in the Gränjen parish and which takes its common name from the Pleiades produces, as we have noted, iron that is cold short in the highest degree — to such a degree that its use has been prohibited. These ores are seen to be composed of two materials, one granular [endowed with a metallic brilliance (*2nd ed.*)] and resisting the attraction of the magnet, the other spathic [blackish and magnetic (*2nd ed.*)], intermixed in the form of nuclei. We have first examined these separately.

Expt. 206. A hundredweight of the granular material, calcined in the usual way on a little plate, gave off no odor of arsenic, not even when burning charcoal dust (which elicits arsenic most efficaciously from others) was added to it. To avoid error, the charcoal dust was previously carefully examined, for it often happens that the charcoal dust exposed by itself to fire emits an unmistakable odor of garlic.

Expt. 207. The same result occurred when in place of the grainy kind the spathic mineral was subjected to the test.

Expt. 208. We also tested in the same way the Gränjen wrought iron itself, reduced to filings, but which gave no sign of arsenic.

Therefore arsenic does not cause this brittleness: moreover, evidences of it only rarely [*2nd ed.*] occur in full-scale operations. Moreover wherever arsenic is present in iron in which this fault flourishes, it hinders ductility not only in cold iron but also in heated iron.

(C) We do not deny that zinc is present in certain iron ores, as at Aix-la-Chapelle and elsewhere, but we do not dare affirm, with certain rather recent authors, that this semimetal is the general cause of cold shortness. [52] For not only is zinc united to molten iron with difficulty and even when that has been done it freely flies away again, but experience testifies that iron does not always become cold short where the zinc mineral abounds, and both the color of the flames and the crust of cadmia in the furnace mouth bear witness to an abundance of the same.[18]

No traces of zinc appear in those Swedish furnaces that are charged with ores that produce cold-short iron, although even when present in traces, zinc can be hidden only with great difficulty. Moreover Experiments 247 and 248 establish it beyond doubt that no amount of zinc can be found in those types of iron that are cold short in the highest degree.

(D) We have collected the residua of the iron that are insoluble in vitriolic acid, washed them properly and dried them out, and then examined their weight and properties. In order that the residuum might be more apparent, we tested two or three hundred-weight of this iron in this solvent, but for the sake of an easier comparison in calculation, we have reduced the results to parts per hundred. Here is the outcome of this work.

---

[18] The Honorable Stockenstrőm has noted such examples for us. In the iron work-shop near Aix-la-Chapelle he saw hammers cast for forging iron which could be kept in use for four months, but in the area around Luxembourg from similar mineral contaminated with zinc only the cold-short type could be extracted. [*Grignon rightly adds the objection that a brittle wrought iron is not necessarily produced from a brittle pig.*]

[Experiment no.]	[Kind of iron]	[Weight of residuum] assay lbs.
Expt. 209.	Leufstad cast iron, saturated with phlogiston, from a hundredweight, left	4.0
Expt. 210.	The same, endowed with a moderate amount of phlogiston	3.3
Expt. 211.	The same iron, wrought	0.3
Expt. 212.	Wrought iron from Österby converted into steel	0.6
Expt. 213.	The same merely wrought	0.1
[53]		
Expt. 214.	Ullfors cast iron	2.0
Expt. 215.	Wrought iron from Ullfors	0.1
Expt. 216.	Cast iron from Åkerby	2.6
Expt. 217.	Wrought iron from Åkerby	0.5
Expt. 218.	Cast iron from Forsmark	3.0
Expt. 219.	The same converted into steel	0.5
Expt. 220.	Wrought iron from Forsmark	0.1
Expt. 221.	Cast iron from Hällefors, loaded with phlogiston	5.3
Expt. 222.	The same, poor in phlogiston	4.3
Expt. 223.	English steel	0.4
Expt. 224.	Steel made by Dr. Quist	0.6
Expt. 225.	Native Siberian iron	0.1
Expt. 226.	Brattefors cast iron	2.5
Expt. 227.	Wrought iron, Brattefors	0.3
Expt. 228.	Steel from Dalia iron	0.5
Expt. 229.	Cold-short wrought iron from Norrberg	1.5
Expt. 230.	Cast iron from Husaby, brittle when cold	6.7
Expt. 231.	The same converted to steel	1.7
Expt. 232.	The same, wrought	0.6
Expt. 233.	Wrought iron from Gränjen, brittle when cold	0.1

From these experiments it is quite clear that cast iron produces the greatest residuum and indeed that cast iron which is saturated with phlogiston produces more than that which is poor in phlogiston; steel leaves far less; but wrought iron leaves a very little and sometimes almost nothing.

The residuum of cast iron has the appearance of plumbago, scaley, slippery, dark, and staining.

Expt. 234. A hundredweight of this residuum, placed on a little plate in a red-hot muffle of an assay furnace, quickly turns white and loses more or less of its weight, but the loss never [54] exceeds half. On examination the dust that remains was found to be of a siliceous nature.

Expt. 235. A hundredweight of the same residuum detonates with niter and alkalizes about five times its weight. That is, if the siliceous matter is subtracted, the remainder needs ten times its weight of niter to deprive it of phlogiston. When this operation is performed in a pneumatic apparatus, about 50 cubic inches of aerial acid was evolved.

Expt. 236. Half a hundredweight of residuum was neither dissolved nor changed in any way by boiling in muriatic acid.

Expt. 237. A hundredweight of the same residuum mixed with an equal weight of iron calx and exposed for 15 minutes in a closed crucible to a fire driven by a bellows gave a regulus of 30 pounds, as also happened with common plumbago.

If one omits the siliceous part, the nature of the residuum of cast iron revealed by these experiments identifies plumbago, for it agrees in all respects with the common variety, even to the point that it is composed of aerial acid and phlogiston.[19]

The residuum from a hundredweight of steel rarely exceeds half a pound and never amounts to a whole pound except for that which is prepared from cold-short iron.

In order that we may truly understand the nature of this residuum, we have first examined that which remains after English

---

[19] Cronstedt in his *Mineralogy*, Sect. 154, lists under the same genus two minerals that differ a great deal in their composition. The first, designated by the letter "*a*," is lamellar, the metallic earth in which it is united with common sulphur: this is distinguished by the name of *Molybdena*. Next is the species of sulphur combined with aerial acid and phlogiston: this is called *Plumbago*. Cf. *K. Vetenskaps Acad. Handlingar* (Stockholm) for 1778, 1779, and 1781. [*Despite their totally different chemical natures, molybdenite (MoS_2) and graphite are remarkably similar in appearance and physical feel, and were often confused under the name "molybdena" — a name which, to further the confusion, was originally applied to lead compounds. Cronstedt first called attention to the distinction between the two minerals, and Scheele, in the papers referred to, identified the true character of molybdenite and prepared the oxide of its metallic base.*]

steel has been dissolved. Nine hundredweight afforded only 3.6 pounds, which in no manner sufficed for correctly measuring the capabilities but nonetheless disclosed its nature. [55]

Expt. 238. A small sample, calcined on a little plate under the muffle of an assay furnace, left a white siliceous residuum.

Expt. 239. When the small sample was added to five times its weight of niter fused in a crucible, it detonated violently, but something blackish remained, which indicated that a greater quantity of niter was needed. We tested if the remainder could produce detonation when added to [*more*] niter, but in vain.

Expt. 240. A small sample was not dissolved in muriatic acid.

These properties are more than enough to demonstrate the plumbagolike nature of the residuum, perchance completely agreeing with the common [*plumbago*] with respect to the nature and proportion of its constituent parts, although we do not dare to assert it strongly since the paucity of the material did not permit a proper exploration.

With regard to that which pertains to the residuum of wrought iron, this is so slight that we were unable to collect from a few hundredweight the amount necessary for examination.

Hot-short iron gave a very large portion, the nature of which we have examined by three experiments.

Expt. 241. The small sample detonated when plunged into red-hot niter.

Expt. 242. Another part, on being calcined, left a white siliceous residuum.

Expt. 243. Another part, boiled in muriatic acid, underwent no change.

[Once these things have been recognized, we can make firm the conclusion that was previously proposed conditionally about the amount of phlogiston in the different states of iron (Section 4, G). That is to say, cast iron does indeed bring forth less inflammable air than does wrought iron, but since it contains more plumbago, the suspicion easily arises that cast iron would have given off the same or a greater amount had the place of the plumbago that was abstracted been occupied by an equal weight of that kind of iron that is present in cast iron. An example may serve to illustrate. Phlogiston-saturated Leufstad cast iron contains 4 pounds of plumbago in a hundredweight (Experiment 209), which are not decomposed during solution (Experiment 236). Therefore in the experiment set up in Section 4, only 100 − 4 = 96 pounds produced inflammable air, but in the wrought iron used 100 − 0.3 = 99.7 pounds were soluble (Experiment 211). Hence it appears that the unequal weights offered only a lame comparison. Therefore let us see it more accurately. Since 96 pounds (of iron in cast iron) produced 43 (cubic inches of inflammable air) (Experiment 1), 100 necessarily will give about 44.8; similarly, while 99.7 (pounds of iron in wrought iron) produces 50 (Experiment 3), 100 will afford 50.1, it manifestly appears that the abundance of phlogiston is greater in wrought iron. The remaining experiments, examined in the same manner, strengthen the conclusions elicited in Section 4. (*This paragraph in 2nd ed. only.*)]

(E) Finally, the acid of sulphur is commonly advanced as the cause of hot shortness. Iron that labors under such a fault to a very notable degree can be provided only with difficulty, since under the hammer it is easily distinguished by its cracks and in vain is it offered to a buyer. Nevertheless we were permitted to examine such a variety from the Norrberg parish, and at the same time both ductile and cold-short irons were exposed to trial. Consider the results. [56]

Expt. 244. Hot-short Norrberg iron was dissolved in the purest muriatic acid, to which were then added some drops of salified heavy earth[dd] dissolved in distilled water, but no precipitate of any sort was formed in the liquid, nor was anything deposited

---

[dd] Terra ponderosa salita; *supposedly barium chloride made from the natural carbonate by treatment with hydrochloric acid. Bergman would have found the sulphur in this hot-short iron had he dissolved it in nitric acid, for in hydrochloric acid most of the sulphur passes off in the gas as H_2S.*

within an entire day, which would undoubtedly have happened had there been present even a quarter of a grain [*of sulphur*].

Expt. 245. The same test set up with cold-short Braås iron had the same result.

Expt. 246. Good ductile Österby iron similarly behaved itself when the same test was repeated with it.

(F) In order that we might understand if perchance any other foreign materials might contaminate cold-short and hot-short iron, we have attacked the problem in the following manner.

Expt. 247. We dissolved a hundredweight of cold-short Braås iron in purified nitric acid, precipitated the solution with the purest alkali from tartar [*i.e., potassium carbonate*], dried out the sediment after it had been carefully washed, and finally exposed it in a crucible to strong fire. This calx then was immersed in nitric acid and was left quiet for some days at a moderate temperature in order that anything soluble might be taken up, but nothing was precipitated [*from the resulting solution*] either by alkali from tartar or by phlogisticated alkali [*potassium ferrocyanide*] except a portion of the iron calx.[20] We have therefore concluded that no foreign metallic materials or earths that are soluble in nitric acid were present.

After the precipitation had been performed, we crystallized the remaining liquid and collected the niter formed from the vegetable alkali. Later, having saturated [*the alkali*] by the action of vinegar, we separated the two crystals in spirits of wine. When the test was performed, we were then able to run down only the [*two*] salts, niter and acetated vegetable alkali, each of which seemed to be pure. And so no traces of foreign acids were noted by this method.[21]

---

[20] We then still believed that the white calx was ferruginous.[ee] [*Footnote in 2nd ed. only.*]

[21] In Dissertation XXVII we show that a foreign metal is present. [(*Footnote in 2nd ed. only.*) *See also the discussion of the white precipitate formed from cold-short iron when dissolved in sulphuric acid, Expts. 264-268.*]

[ee] *The precipitate was, of course, iron phosphate. Bergman examines this in detail in a paper published later in 1783, referred to in the next footnote.*

Expt. 248. We treated in the same way a hundredweight of hot-short iron. The precipitate was dried out and ignited for a long time, then [57] mixed with nitric acid, but we detected no soluble foreign matter, neither in the precipitated calx nor in the liquid remaining after precipitation.

Expt. 249. We precipitated Husaby cast iron, dissolved by acid, by the action of phlogisticated alkali. Two hundred and twenty-eight pounds of the precipitate, after drying, placed in a crucible with a layer of charcoal dust and sealed, gave [*after heating*] a brittle regulus of $39\frac{1}{2}$ pounds. This regulus, when mixed with a quarter [*of its weight*] of calx of good iron, retained its original weight and the nature of cold-short iron [*2nd ed.*].

Section 9
The Proximate Principles of Iron

If those things that have previously been said in detail are clear, it will be obvious that iron, whose nature is different in so many ways from that of other metals, nevertheless generally agrees with them in respect to composition to such an extent that one may perceive only two proximate principles, a metallic earth that is truly peculiar to it, and phlogiston.

(A) Moreover, since inflammable air can be obtained from any type of iron, and indeed in the same volume even when different acids are used, a suspicion could perhaps arise concerning this gas [*aura*] hiding in the metal itself, in much the same way as calcareous rocks and various other materials support aerial acid. But if this suspicion were well founded, every acid attacking iron ought to produce inflammable air, but experience testifies to the contrary. For acid of arsenic dissolves iron without the production of any elastic fluid, for the phlogiston about to escape is instead joined with a portion of the solvent, regenerating white arsenic. Moreover nitric acid produces no inflammable air but produces some gas known by the name of nitrous air. To be sure it may be retorted that this acid attacks inflammable things [58] with such violence that perchance the inflammable air itself is destroyed in the solution, but it has been proved in other experiments that very strong nitric acid has no power whatever to change inflammable air, which therefore is not evolved or freed but in truth is born in the solution. [Moreover, inflammable air is evolved from iron by other methods, certainly by the action of water (Section 10), or it can be extracted even by fire alone. Its genesis in the former case will soon be examined (Experiment 257), and its genesis in the latter case by the phlogiston of metal combining with caloric matter is discussed elsewhere.[22] (*2nd ed.*)]

Furthermore, preparatory to exploring whether all inflammable air is of the same nature, we have collected pure inflammable air in equal bottles which hold 6 cubic inches for examination.

Expt. 250. Inflammable air from hot-short Norrberg wrought iron, when a burning candle was applied to the opening, caught fire with a certain explosion and shining forth with diffuse flames, consuming the entire store in a moment. When the candle was applied again, no signs of inflammation appeared.

Expt. 251. Air from cold-short Braås wrought iron behaved similarly, but after the first conflagration had suddenly passed, the remainder was still combustible, burning for about two minutes, at first with weak yellowish flames which turned slightly blue around the base of the mouth of the bottle.

Expt. 252. Air from Leufstad cast iron endowed with a moderate amount of phlogiston excited yellowish flames with the explosion which, as the others, were quickly extinguished, but when the candle was applied again a new flame was produced, burning weakly for two minutes, turning a little red and sparkling.

Expt. 253. Air from Leufstad cast iron behaved in the same manner, but the second flame turned more toward yellow.

Expt. 254. The air collected from English steel at first caught

---

[22] *De Attractionibus Electivis Disquisitia*, Sect. 39. [*In Bergman's* Opuscula, *1783, Vol. III, pp. 291–470.*]

fire like that from Leufstad wrought iron, but the second flame, lasting for about two minutes, was weak and blue.

These experiments evince some differences, which seem especially notable in the case of hot-short iron. [59]

(B) The variations of iron owe their own origins not only to the proportions of the principles but often also to other materials which are not necessarily always present but may be, and in fact sometimes are, absent. But although this is the case, these foreign things nevertheless often are related to the different states of our metal, for when the materials are removed, the states also are profoundly changed. And so perhaps it will not be useless to present here a glance, albeit a limited one, according to the experiments that have been performed. We have added at the end the specific gravities of the varieties previously examined [although it is necessary to confess that these are not very trustworthy because of the great difficulty in excluding air from the rough surfaces (*2nd ed.*)].

CAST IRON

	Min.	Max.
Siliceum, in a hundredweight,	1.0	3.4 lbs.
Plumbago	1.0	3.3
Manganese	0.5	30.0
Iron	63.3[ff]	97.5[ff]

to which as much phlogiston is present as afforded 38 to 48 cubic inches of inflammable air.

The combined caloric matter[gg] corresponds to 20 to 52 degrees.

The specific gravity of Leufstad cast iron saturated with

[ff] *The minimum and maximum figures of iron in this and the following four tables are reversed in the 1781 printing, and the columns are not headed "Minimum" and "Maximum." Some other errors have been corrected in accordance with the list of corrigenda at the end of the 1781 essay.*

[gg] *The second edition reads:* "The amount of combined caloric matter inherent in steel and ductile iron is inversely proportional to the degrees increase in temperature with nitric acid, that is, the minimum can be expressed relatively as 1/52 and the maximum as 1/20." *Bergman seems then to have thought that the more heat was combined with the iron the less would be available for release when it was dissolved.*

phlogiston is 8.062 [*second edition reads 7.662*]; of that endowed with a moderate amount of phlogiston, 7.759.

(C)   STEEL

	Min.	Max.
Siliceum, in a hundredweight,	0.3	0.9 lbs.
Plumbago	0.2	0.8
Manganese	0.5	30.0
Iron	68.3	99.0

bearing as much phlogiston as 44 to 48 cubic inches of inflammable air.

The combined caloric matter corresponds to 74 to 114 degrees.[hh]

The specific gravity of Husaby steel is 7.002; of that prepared by Dr. Quist, 7.643; of English steel, 7.775; of Forsmark, 7.727; of Österby, 7.784; and of the same hardened, 7.693. (All the other samples of steel had been softened.) [60]

(D)   DUCTILE WROUGHT IRON

	Min.	Max.
Siliceum, in a hundredweight,	0.05	0.3 lbs.
Plumbago	0.05	0.2
Manganese	0.50	30.0
Iron	69.5	99.4
	[or 99.4	99.5 (*2nd ed.*)]

and the same quantity of phlogiston as 48 to 51 cubic inches of inflammable air.

The combined caloric matter corresponds to 122 to 136 degrees.[ii]

The specific gravity of Brattefors wrought iron, 7.798; of Leufstad, 7.754; of Österby, 7.827.

(E)   HOT-SHORT WROUGHT IRON

Siliceum, in a hundredweight,	0.8
Plumbago	0.7
Manganese	0.5
Iron	98.0

endowed with the same amount of inflammable matter as enjoyed by 48 cubic inches of inflammable air.

[hh] *The second edition reads:* "The relative amount of the combined caloric matter falls between the limits 1/114 and 1/74."

[ii] *The second edition reads:* "The combined caloric matter may be expressed as a minimum by 1/136, as a maximum by 1/122."

The combined caloric matter corresponds to 130 degrees[jj].
The specific gravity of hot-short Norrberg iron is 7.753.

(F)   COLD-SHORT WROUGHT IRON

	Min.	Max.
Siliceum, in a hundredweight,	0.05 [0.05][kk]	0.3 [0.5][kk]
Plumbago	0.05 [0.05]	0.3 [0.4]
Manganese	0.5 [0.50]	4.0 [4.0]
Iron	95.4 [95.40]	99.4 [99.4]

the reducing phlogiston in it produced as much as contained in 50 to 52 cubic inches of inflammable air.

The combined caloric matter corresponds to 122 to 134 degrees.[11]

The specific gravity of cold-short Braås iron, 7.792; of the same iron but ductile, 7.751; of Husaby, 7.791. [61]

(G) Now in the last place it ought to be explained what effect any material ingredient may have in shaping the qualities of iron, but very many experiments still remain to be performed in order that this may be done with full success. Meanwhile, it may suffice here to touch lightly upon only the more notable points. Let us first see whence the three states of iron arise.

In cast iron the individual heterogeneous substances are each present in a larger quantity than in the other [*forms of iron*], whence the assemblage is necessarily more brittle. The amount of phlogiston varies a little, so that in that type in which it is more abundant, plumbago abounds proportionately, which seems at the same time to promote softness and to retard fusion. We have previously said that plumbago impedes mineralization, that the iron particles do not receive phlogiston to complete satiety, but our experiments at first sight seem to disagree with this thesis since they testify that cast iron saturated with phlogiston also abounds with plumbago. But it is necessary to consider that cast iron that is saturated with

---

[jj] *The second edition reads:* "may be signified as 1/130."

[kk] *The figures in brackets are those of the second edition.*

[11] *The second edition reads:* "is indicated as a minimum by 1/134, as a maximum by 1/122."

phlogiston is nevertheless always inferior to steel and wrought iron in its own genus with respect to the abundance [*of phlogiston*], and moreover the tests that we have advanced give a signal, as it were, that the phlogiston that is taken up in this case is distributed in a definite proportion between aerial acid and iron molecules. Siliceum enters as an admixture but is not dissolved. The many iron molecules that can be contained in this state, provided with a dose of the inflammable principle, resist the fire better, but if they are deprived [*of phlogiston*] they are more easily calcined and pass into the slag with the siliceum contained in it.^{mm}

The state of steel is closely related to cast iron. Indeed in this state heterogeneities are far more scarce than in cast iron but more

---

^{mm} *In the second edition this paragraph reads as follows:* "In cast iron individual heterogeneous substances are each present in a larger quantity than in the other [*forms of iron*] whence the assemblage is necessarily more brittle. The amount of phlogiston varies a little, so that in that type in which it is more abundant, plumbago abounds proportionately. A small dose of plumbago produces white cast iron — shining, heavy, hard, and easily melted, but losing 20 to 24 parts per hundred by weight in the forge. But cast iron with a more or less blackish color [because it is loaded with a grayish plumbago] when deprived of this becomes lighter and softer and flows with far more difficulty, but when processed in the forge it suffers a loss in weight of only 10 to 12 parts per hundred. For the great amount of iron molecules that can be contained in this state, that is, provided with an abundance of inflammable principle, resist the fire better but when deprived [*of phlogiston*] are more easily calcined and pass into the slag with the siliceum that is present. The Gallic and Osmund methods of forging require cast iron having a moderate amount of plumbago, and they generate a richer slag which helps in the destruction of the plumbago. But for the German and rustic methods, aimed not so much at the quality as at the quantity of the iron, cast iron saturated with phlogiston is more appropriate. We have previously said that plumbago impedes mineralization, that the iron particles do not receive phlogiston to complete satiety, but our experiments at first sight seem to disagree with this thesis, since they testify that cast iron saturated with phlogiston nevertheless also abounds with plumbago. But it is necessary to consider that cast iron that is saturated with phlogiston nevertheless is inferior to steel and wrought iron in their own genera, and moreover the tests that we have advanced give a signal, as it were, that the phlogiston that is taken up in this case is distributed in a definite proportion between aerial acid and the iron molecules. Siliceum enters as an admixture but is not dissolved."

abundant than in ductile iron. Steel is also richer than cast iron in the inflammable principle and in hidden heat.[nn] Plumbago seems to be necessary to each state but in a definite portion for each. Because of it, undoubtedly, both cast iron and steel are darkened by acids [62] (especially by nitric acid), for when the metallicity has been eaten away, a very thin and dark web of plumbago remains on the surface. In addition, the hardness that is attained when a metal more or less heated is suddenly plunged into water also seems to depend in some manner on a moderate dose of the same plumbago, for ductile iron increases its hardness only a little by this operation, cast iron a little more, and steel a great deal. The mechanism by which this is accomplished remains to be unearthed by further scrutiny. It does indeed seem fairly reasonable that a body that has been heated hardens when it is suddenly made cold, and its particles, momentarily inflated by fire, are fixed as it were, for the specific gravity is less in hardened than in soft steel, which signifies the expansion of the volume. But the manner in which half a hundredth part of plumbago is able to produce so great a difference is a Gordian knot not easily unraveled.

Elasticity generally seems to depend upon the inflammable principle joined in a definite amount with combined caloric matter. At least its genesis in certain cases supports this hypothesis. Thus, boiling acid of vitriol begets elastic vapors, but when they are made cold again they condense into a liquid, but if a little inflammable matter is dropped into boiling acid, it soon produces a copious elastic and permanent gas. The same applies to nitric acid which indeed, since it has been deprived by the boiling of the portion of inflammable principle that is accustomed to be welcomed into the same, then being without phlogiston and heat emits no permanent air.[oo] Muriatic acid always produces such air

---

[nn] *The second editions says steel* "lacks the abundance of combined heat."
[oo] *Latin unclear. It apparently means that the dephlogisticating (i.e., oxydizing) property of the acid prevents the formation of inflammable air (i.e., hydrogen). The volatile portion is identified as the inflammable principle only in the second edition.*

by the action of heat, for it nourishes inflammable air within its structure. Steel, although pre-eminent with respect to elasticity, is nevertheless always inferior to ductile iron in both the quantity of inflammable principle and of combined caloric matter [*second edition reads:* "is inferior to ductile iron in the amount of inflammable principle but exceeds it in the amount of combined caloric matter"], whence we learn that a balanced proportion is more suitable for producing a very great elasticity in this metal than is accumulation without limit. In all nature causes have been similarly circumscribed and determined [to such a degree that a very great effect may be produced by a more or less weak predetermined force or addition. Thus the resolution of water into steam is accelerated the more as it is tormented by a more intense heat up to +150 degrees, where the operation seems to be completed with the least delay. This temperature is like the point of inflection in the higher Geometry, for the evaporation is progressively retarded as the heat is further increased].[pp] [63]

The surface of a polished steel sheet that is gradually heated acquires different colors in accordance with the degree of heat, undoubtedly from a slight but increasing calcination of its outermost molecules.

Ductile iron can properly be called pure, and although it is very greatly enriched with phlogiston [and bound-in caloric matter (*1st ed. only*)], it nevertheless resists fusion with far more pertinacity than the preceding varieties of iron. This again shows clearly how important the proportions are. [For we have observed at the beginning of this Section G that fusibility is increased by a small amount of plumbago but is decreased by a copious amount of the same. (*2nd ed.*)] The same is true regarding ductility, which increases with the phlogiston [and bound-in caloric matter (*1st ed. only*)] in the different states, but when that is increased beyond a

---

[pp] *The material in brackets is in the second edition only. The temperature refers to that of a hot surface on which the water is dropped. When it is above 150°C the drop is insulated by a layer of steam, and the rate of evaporation decreases.*

certain limit, fragility appears in cold-short iron which, having been made malleable by the action of limestone (Experiment 31), produces only 48 cubic inches of inflammable air in place of 52 (Experiment 32). Moreover, it is noteworthy that the caloric matter does not vary here, although the nature of the iron has been changed (Experiments 165, 167).

[And so the various states of iron have been derived from the diverse unions of phlogiston, caloric matter, and plumbago. But there is some latitude to the limits of whichever state may be chosen in the numerous differences that may occur (depending on the differing proportions of the three materials just noted) in color, specific gravity, ductility, hardness, elasticity, and tenacity. Also by its very mechanics the handling of iron gives rise to some discrimination. The particles are compressed by long-continued forging, which increases the specific weight and produces a greater rigidity of structure as well as those things that are mutations of the others. (*2nd ed. only*)]

Let us now consider more carefully the nature of defective iron. Hot-short iron can be extended, filed, and bent especially when it is cold and generally even when it is heated white hot, but when it is heated red hot it is broken. The fracture is fibrous. If the surface is filed, it easily becomes blue and rusts. Pots that are cast from this iron darken food and give out the smell of the furnace. [*Forged*] at a white heat, this iron hurls out sparks more frequently and of larger size than comes from good iron. The fault is not corrected by its opposite, cold-short iron (Experiment 125), nor by calcination (Experiment 129), but it seems to be improved a little by cementation (Experiments 138–140), although we must admit that the Norrberg iron is less defective [*than most irons*]. The whole nature [*of hot-short iron*] should not be attributed to acid of sulphur (Experiment 244) nor to the proportion of heterogeneous things or its principles, but it seems it is chiefly derived from the quality of the phlogiston having been defiled in some manner. These things give a sign: the different character of the flame of the air begotten from this (Experiment 250), its behavior with sulphur and water [64] (Experiment 261), its odor, its sparks, etc. It also has been noted that good iron becomes hot short when it is worked with poorly burned charcoal. [Here the following observations due to Dr. Stockenström appear most noteworthy. That is to say, if slag that has arisen in the furnace at the forge producing iron that is hot short but not notably so is reduced by resmelting, the defect increases in iron taken from the slag that arose in the first operation but is corrected if the slag that has previously suffered no reduction is added to the iron in the forge hearth. Therefore such slag seems to absorb the cause of hot shortness by the power of a greater attraction, whence also it is understood why cast iron that is easily smelted and produces sufficient slag scarcely ever produces iron spoiled by the defect of hot shortness. Moreover this fault is corrected with more difficulty in the forge hearth than in the smelting furnace, where the metal attains complete liquidity. (*2nd ed.*)]

Cold-short iron is malleable when heated both red-hot and white-hot, it appears silvery white when it has been filed, and it resists rust well. The fracture is full of grains [that are white and shining (*2nd ed.*)]. It is not corrected by being worked in the forge with a quarter of its weight of calx of good iron (Experiment 100), and with half its weight it seems to acquire hot shortness (Experiment 101). Cast iron of this nature [sometimes (*2nd ed.*)] shows a wonderful volatility with calx of iron (Experiments 102–104). It is corrected with limestone in a Hungarian furnace (p. 183, fn. 4), but not in a crucible (Experiments 120–123). When treated with sulphur it passes into steel (Experiment 124) and similarly when it has been calcined and then melted with good iron (Experiment 128). It is not changed by cementation (Experiments 141–143). Turned into Prussian blue and afterwards reduced, it keeps its own nature (Experiment 249). The origin of these qualities seems to be deeply buried. Without a doubt[qq] the proportion of phlogiston has

[qq] *The second edition reads* "Perchance."

some effect (Experiments 31, 32), but unless we err, the principal foundation lies hidden in the portion that is present of ferruginous calx of a peculiar nature and will soon be examined (Experiments 264–266).

Section 10
On the Calx of Iron

Thus far we have sought the proximate principles of iron and have briefly noted how we have arrived at their amounts by decomposition of its calx of the same. A thing that indeed seems to merit attention from the outset is the peculiar power of water on our metal. Concentrated acid of vitriol does not dissolve iron unless water has been added to it, and even alone water reduces iron to a powder [of *ethiops martialis* (*2nd ed.*)] if it is immersed for a long time, and it also aids the action of sulphur on iron filings so greatly that not only does a vehement heat arise, but even smoke and flames appear. Do not these phenomena in some manner indicate that there is a radial salt, [65] hidden especially in acid, that agitates water? But let us examine these things more properly.

Expt. 255. Eight hundredweight of filings of Österby iron, moistened with some drops of distilled water, expanded the mercury of an immersed thermometer from 14 degrees to 17 degrees which, without a doubt, proves some reciprocal action.

Expt. 256. Three hundredweight of the same filings on a small porcelain plate were covered with distilled water, which gradually evaporated at ordinary temperature. When it had dried, the mass was ground with an agate pestle, and water was once more added; the operations were continued thus for 60 days. After this time, the ferruginous particles that had been resolved into a black dust were still strongly attracted by the magnet, but some that had turned into yellow ochre were indifferent to it. The weight of the whole now was 371 pounds, which indicates an increase of 23.7 pounds in a hundredweight.

Expt. 257. One hundred and eighty-five assay pounds of the same filings with 688 of distilled water were immersed in a small cucurbit of capacity equal to 3200 [*assay*] pounds of water or, when empty, 4 cubic inches. A slender little tube was adapted to fit perfectly all around the orifice, bent so that its free end might enter another more ample tube filled with mercury, and inverted with its opening immersed in mercury. This apparatus was left at average temperature. On the third day the mercury began to rise in the little tube and filled it all the way up to 2/3, but the following day after various oscillations it occupied only 1/3 and remained fixed there for two months. Meanwhile, in fact, some calcination was observed in the form of yellow stains. When the stated time had elapsed, the water was strained through a paper filter and examined: it did not change the color of [*the indicator*] turnsole and produced no iron [*reaction*] with additions of either some drops of phlogisticated alkali or some drops of tincture of gallnuts. The filings, [66] when collected and dried, weighed 182 pounds, and the black dust, which was very sensitive to the magnet, together with some ocherous scrapings, weighed about three pounds. Some ocher stains remained fixed in the small cucurbit.

It appears from this experiment that water is not inert even in a closed place, although the air was at the same time diminished by about 1/500, for the initial volume of air was equivalent to 2528 pounds of water and the final, to 2523 pounds. [The lack of access of any free atmospheric air undoubtedly placed an obstacle to further calcination. For it has been noted that filings of iron immersed in water produce some inflammable air and gradually drop down into a black dust, which although all of it is magnetic, nevertheless in an acid produces not more than 3 (cubic) inches of inflammable air from a hundredweight (Experiment 272). (*2nd ed.*)]

Expt. 258. Thirty-two hundredweight of phlogiston-saturated Hällefors cast iron, mixed with the same amount of sulphur and 4 hundredweight of hot water in a little porcelain flask, grew hot and gave off a visible smoke after 8 minutes, ceasing after 45 minutes. The surface became darkened.

Expt. 259. Thirty-two hundredweight of Husaby cast iron when similarly treated behaved in the same manner, with a little greater heat.

Expt. 260. An equal weight of Österby wrought iron produced visible smoke after 15 minutes, continuing for an hour and a quarter. The greatest heat occurred after 45 minutes had elapsed. When cold, the mass was a darkened yellow color.

Expt. 261. Thirty-two hundredweight of hot-short Norrberg wrought iron after 12 minutes sent up a perceptible smoke that was visible for more than an hour and a quarter, although the greatest heat began to decrease shortly after 45 minutes. The outside of the mass was blackish.

Expt. 262. The same weight of cold-short Braås wrought iron showed the first smoke after half an hour, continuing for an hour and a quarter, with the greatest heat evolved after three-quarters. The mass was a darkening yellow color.

These experiments were set up simultaneously in the same place. Cracks opened up in the mass in each of them, but no flame shot out for this requires at least a pound of iron. [67] Some degrees of difference did indeed occur here but our plan did not make progress, and so we have broken this web and tried another way.

(B) We know that when wrought iron is dissolved in vitriolic acid it produces vitriol in which iron molecules have been deprived of the reducing inflammable principle. When this vitriol has been dissolved again and exposed to the air in an ample vessel, we see that the clear liquid is quickly clouded by yellow particles, which turbidity does not cease before the whole liquid, being reluctant to crystallize, has been resolved into a lye. The reason for this has been set out elsewhere.[23] This provides an enlightening example

---

[23] *Opuscula*, Vol. I, p. 172.

of complete dephlogistication. The following experiment shows what happens in a closed jar aided by a boiling heat.

Expt. 263. The flask in the apparatus previously described (Section 3, A) was filled with 4 hundredweight of well-crystallized iron vitriol dissolved in water. On exposure to a heat of 100 degrees, about half of the liquid was got rid of, and after this the other half boiled for a quarter of an hour, although it evolved no more air than is customarily present in distilled water, and this air could not be set afire in any way. The liquid was turbid, and great amounts of ochre were deposited when its movement ceased. This yellow sediment could be separated even at a heat of 70 or 80 degrees without boiling, and in this case only a little of the solution was driven off.

Hence it is sufficiently apparent that dephlogistication of iron occurs by heat alone without the help of air and produces no inflammable gas.

It has been sufficiently noted how good iron may behave when dissolved in an acid and then precipitated by an alkali. A solution of hot-short iron acts similarly, but a solution of cold-short iron shows peculiar phenomena. [68]

Expt. 264. Cold-short wrought iron was dissolved to saturation in vitriolic acid. The solution, properly diluted, was exposed to the air in an open jar, and after some hours this spontaneously began to become turbid with snowy particles, which gradually settled.[rr] The first sediment was collected, and then another, which was yellowish, followed. The separation was accelerated by the action of heat.

Expt. 265. This sediment remains snowy white when properly washed and dried. It acts [*on analysis*] in the humid way thus: it is

---

[rr] *This white precipitate was probably ferric phosphate, which is relatively insoluble. Here Bergman definitely and correctly associated hot shortness with the presence of something that would form this precipitate; the identification of this something with phosphorus was made by J. C. F. Meyer in 1784. Bergman reported more comprehensive studies in 1783. (See the second footnote to Expt. 268.)*

dissolved without effervescence by acids — vitriolic, nitric, muriatic, arsenic, and perchance by all others — and it was precipitated again in a white form by fixed aerated alkali, but phlogisticated alkali gave blue. It was scarcely dissolved either by fixed or by volatile alkali unless it had been recently separated and was still moist.

Expt. 266. This [*white precipitate*] when explored by the dry method shows the following behavior. It does not lose its whiteness by burning. It is as indifferent to the influence of the magnet afterward as it was before. When exposed on charcoal to the action of an ironworker's tube [*blowpipe?*], it becomes ashen in color and is afterward scarcely soluble in acids. After it had been melted, it was taken up by microcosmic salt and by borax, showing darkened glassy globules the same as with the calces of good iron and of hot-short iron. If niter is added to the white sediment in a crucible, it boils up and reddish smoke immediately arises. If nitric acid is often removed by ignition, the white color is changed by darkening with red.

Expt. 267. In order that we might determine the quantity of good calx, we reconverted into crystals by evaporation 255 pounds of the same cold-short iron dissolved in vitriolic acid. We obtained 938 pounds of crystals. We precipitated these by fixed alkali and collected the rust-colored calx which weighed, after drying, 288 pounds. When reduced in a crucible with a layer of charcoal dust, this gave a regulus of 167 pounds, which, by remelting with a quarter of its weight of good iron calx, acquired a wonderful ductility and weighed 181 pounds.

Expt. 268. The sediment [69] collected under the preceding process from the remaining lye, which spurned crystallization, yielded 96 pounds after evaporation to dryness. It was reddened by ignition on account of the presence of a portion of good calx. After reduction it gave a regulus of 52 pounds. When this was mixed in the usual manner with calx of good iron, it produced iron that was cold short in the highest degree, with a loss in weight of

15 pounds. Therefore in two reductions 36 pounds [*of the original 255*] went away into slag.^{ss}

By pondering over the four immediately preceding tests it clearly appears that the white calx is the cause of cold shortness. We have concluded from this that this calx contains more phlogiston than the common yellow type or that of a dark reddish color, which after ignition is easily dissolved by acids, even by vinegar and nitric acid (Experiments 205, 247); this sends up red vapors [*when melted*] with niter and treated repeatedly with nitric acid, and ignited it turns red (Experiment 266). Meanwhile, nevertheless, this white calx precipitates under the first evaporation, though [*a large part nevertheless*] remains in the final lye. [*After precipitation with alkali*] it seems to need a larger dose of reducing phlogiston, in order both that it may be retained by acids and be crystallizable and that it may acquire a completely metallic form (Experiments 31, 32). Moreover, we have not yet been able to dig out its structure more exactly on account of the paucity of the experiments.[24]

(C) In order that we might test the dephlogistication of iron calx by the most efficacious means yet known, we acted in the following manner.

Expt. 269. We dissolved a hundredweight of Österby wrought iron in nitric acid in a glass vessel, evaporated it to dryness, then poured in more acid and continued thus alternately until 28

---

[24] *Many things appear about this calx in Commentary XXVII.* [*This note in 2nd ed. only. The commentary referred to is Bergman's famous essay,* De Causa Frigilitatis Ferri Frigida, *"On the cold shortness of Iron," written in 1781 and reprinted on pp. 190–93 in Volume III of the* Opuscula Physica et Chemica, *published in 1783. The contents of this volume were not included in the English translation of the essays.*]

[ss] *In the first edition, Bergman gave the weights in this experiment on the basis of 200 pounds of the precipitated calx and 74 pounds of the final residuum. We follow the second edition, in which he recomputed these results to correspond with the entire initial sample of 255 pounds of iron. The final loss of 36 pounds is the difference between the 167-pound regulus of good iron in Expt. 267 and the 52 pounds in Expt. 268. By the alkaline precipitation in Expt. 267, a pure iron hydroxide would result, but precipitation must have been incomplete, for some remained in solution to be recovered with the phosphorus in the residuum, which would have been mainly potassium sulfate and unreacted potassium hydroxide and carbonate.*

hundredweight of solvent were consumed, and then we dried it out by ignition. Initially, during the vaporization of the first 8 hundredweight, copious red vapors were generated, but thereafter no traces of these appeared. The specific gravity of the acid was 1.268.

The residuum was of a dark reddish color and weighed 139 pounds, [70] an increase of 39 pounds. We exposed this to dephlogisticated acid of salt for 10 days and nights, after which time no change could be discerned, not even a change of color.

Expt. 270. We similarly treated a hundredweight of Norrberg wrought iron with the same acids. The result was the same. The residuum was found to have increased in weight $37\frac{1}{2}$ pounds and resisted dephlogisticated acid of salt the same as the preceding one.

Expt. 271. A hundredweight of cold-short Braås wrought iron, tortured in the same manner, behaved similarly with both nitric acid and dephlogisticated muriatic acid. The weight of the residuum was 140 pounds.

The experiments advanced in the preceding pages sufficiently indicate that it is often possible for iron to lose a little more than its reducing phlogiston (Experiment 263), but not so much by any means heretofore described that it can be deprived of its coagulating phlogiston in order to reveal in some manner a radical acid. But this pertinacity does not deter us; never-ending labor conquers everything. Let us seek with untiring zeal more valid instruments and hope that eventually the radical acids not only of iron but also of the remaining metals will be laid bare.

Section 11
On Magnetism

It has been noted from very ancient times that iron is subject to the sway of a magnet. But it may be doubted whether this alone among the metals submits. Thus far nickel has not been entirely freed from such obedience, indeed rather specially purified nickel exercises magnetic power to such a degree that one half of a regulus[25] attracts the other. Moreover, cobalt and manganese also lose their affinity with the magnet with very great difficulty. [71] We read also that not a few other materials have been similarly tested with a magnet.[26] But we dismiss all these things that will be more accurately explored and hasten on to those that are more in line with our purpose. We know that wrought iron is not only attracted by a magnet but is even able to acquire its force in various manners. We also see that very many of the ores of iron, especially Swedish ores, are very submissive to a magnet. Hence arises the question, which still bothers many people, of whether or not these minerals contain iron that is complete [*i.e., metallic*], identical in nature with wrought iron and to such a degree that it is, in truth, native.

We have used the techniques described in Section 3 A, by means of which this problem also can be solved.

Expt. 272. Martial ethiops, the entire amount of which was magnetic and soluble in acids, when boiled in a flask with acid of salt, produced only three cubic inches of inflammable air from a hundredweight.

Hence it is clearly demonstrated that not only does all iron that can be attracted [*by the magnet*] not correspond to wrought iron but even that is a great deal different from cast iron. And so the name of native iron cannot be applied to all mineral friends of the magnet. Indeed, rather, a natural magnet perhaps never contains complete iron, as may be confirmed by the following test.

Expt. 273. We added just enough linseed oil to martial ethiops so that the mass could be molded into a parallelepiped, and when this had been dried out by a long-lasting and mild heat, we placed it between the poles of the armatures of two natural magnets, and after a certain time we found it to have been imbued with magnetic force.

[25] *Opuscula*, Vol. II, p. 242. [*Eng. trans.*, p. 246.]
[26] A. Burgmans on Magnetism.

We know that ocher when cooked in oil becomes indeed very submissive to a magnet, but neither with nor without effort does it recover in this way the whole [72] complement [*of phlogiston, etc.*] that is necessary to completeness. The same thing happens also to ochre enclosed in glass, chalk, gypsum, charcoal, dust and other materials to impede the fusion of the vessel, surrounded by a vehement fire and exposed for a long time, in which case it drinks in phlogiston from the decomposed caloric matter, by which it becomes magnetic, although it is hardly permissible to hope for perfect reduction, since there was no inflammable matter added that was appropriate for melting.

If in the experiment recently described the oil is saturated with a sufficient amount of lead and the drying-out process is completed in front of a fire at a distance of about six feet, the entire task can be completed in five hours.[27]

Hence it is thought that a certain dose of phlogiston is necessary in order to render iron obedient to the magnet, but a far smaller amount suffices than is needed for the complete metal.

Moreover it now seems to be beyond doubt that magnetic phenomena are provoked by a certain very tenuous fluid, the rules of the movement of which are to be carefully investigated. The tests instituted up to the present show a wonderful analogy and concurrence with electric phenomena,[28] certainly very worthy things that may be pursued further.

---

[27] Consider the method that Dr. Knight had used for composing artificial magnets so strong that when placed on a table they aligned themselves in accordance with the meridian of the place. However, instead of martial ethiops he used a very fine knife-grinder's dust arising from the filings of steel, shaken up in a wooden cylinder with water and consolidated by lengthy and tedious labor. The renowned Wilson has recently divulged the process in *Philosophical Transactions* for the year 1779. [Benjamin Wilson "An account of Dr. (Gawin) Knight's method of making artificial loadstones," Phil. Trans. Royal Soc. *1779, 69,* 51–53. The paper describes an interesting antecedent of today's magnets of high coercive force made from small particles.]

[28] *The Test of the Theory of Electricity and Magnetism* of the celebrated Aepinus. [Franciscus M. U. T. Aepinus, Sermo academicus de similitudine vis electricae atque magneticae, . . . (*St. Petersburg, 1758*).]

So far we have been unable to extract from colcothar of vitriol, exposed to a strong fire, a white earth very prone to a magnet, either by the action of vitriolic acid [73] following Lemery[29] or by the action of nitric acid following [Peter van] Musschenbroek.[30]

These are the facts in our progress in the analysis of iron up to the present. Although they are very imperfect, we hope that they will be neither altogether without use in formulating and confirming true theories nor entirely sterile in illustrating and perfecting industrial practice. Now it still remains to compare our tests with those of others that have generally been set up in a different manner, but we have long ago passed considerably beyond the limits of an academic dissertation, so far in fact that we are forced to stop here, adding only a few points in passing that clearly strengthen the conclusions we have previously brought out. [In considering the numerous experiments of the illustrious Réaumur, we are amazed at his indefatigable patience and felicitous genius in investigation. He rightly estimated the difference of cast iron, steel, and ductile iron from a diverse amount of heterogeneous matter, which he distinguished by the name of "sulphurs and salts." Indeed sulphureous (*i.e., reducing*) matter is in fact present, although not the common vitriolic kind but that now better known by the name of plumbago. We have not, however, been able to get hold of something of the nature of a salt as distinct from a radical. Otherwise his whole theory and almost all his tests are singularly in accord with ours. (*2nd ed.*)] Thus Réaumur by cementing cast iron and steel in *crocus martis* finds that the surface becomes ductile, which we have brought about by fusion so that it occurs throughout the whole mass. This very wise doctor produced a similar, though lesser, softness with chalk, natural lime, fixed alkaline salts, and many other materials [especially with powdered bone ash.[31] That bone

---

[29] *Mémoire de l'Académie des Sciences* (Paris), 1706.
[30] *Dissertatio Physica experimentalis de Magnete* [Leyden (1729)].
[31] *L'art d'adoucir le fer fondu* (Paris, 1722).

ash had been more efficacious than the other materials in decomposing plumbago appears very clearly now that its composition has been recognized. For the acid of phosphorus attacks phlogiston, while lime, as the relict, at the same time seeks marriage with aerial acid. Most of the remaining phenomena are explained with the same ease (*2nd ed.*)]. Dr. Rinman saw that cast iron saturated with phlogiston, after being cemented in limestone powder, is dissolved in nitric acid without a residuum, which proves that plumbago can be decomposed even without melting. He observed that Gränjen wrought iron, indeed, heated continuously by itself for 14 days and nights in well-sealed vessels, formed a calcined crust that was easily separated when struck with a hammer, had lost more than 32/100 of its weight, and had acquired greater fragility than before. He observed that ductile wrought iron treated in the same manner lost more than 22/100 but preserved its nature; steel made by fagoting lost almost 21/100, [74] while that made by cementation lost almost 24/100; good cast iron lost 26/100, and at the same time on its surface it acquired the nature of wrought iron; while cast iron surrounded by powdered bone ash had been softened in the same way on the surface but had sustained a loss in weight of only 15/100 of its weight.[32] We regret that the history of this metal written by our great master in treating iron has not yet appeared, for it undoubtedly will contain a great abundance of observations and experiments of great importance to the unearthing of the nature of iron.[tt]

Very recently the renowned Meyer has also undertaken various experiments illustrating the nature of iron.[uu] He observed that ductile iron when remelted with charcoal dust increased its weight, contrary to our Experiment 107 in which a small decrease was

---

[32] In a Swedish treatise entitled: *Anledning til kunskap om grofare Järn och Stålförädling* [Stockholm, 1772].

[tt] *Sven Rinman's great book,* Försöck till Jarnets Historia, *appeared in 1782.*

[uu] *The following two sentences discussing Meyer's work are omitted in the second edition.*

found, perchance as a consequence of an unusual cause. The surface of our regulus appeared to be burned, and the process was tested only once, while Dr. Meyer saw the consensus of very many tests and used larger masses, so that we do not think anyone should doubt this increase, which is not contrary to our corollaries, until the work can be repeated designedly.

We must now pass by other matters that relate less directly to the sequence of arguments in this little work [and the sound theory of iron (*2nd ed.*)].

CHAPTER NINE

On the Nature of Steel and Its Proximate Principles

Louis Bernard Guyton de Morveau

Extract from "Acier," *Encyclopédie Méthodique. Chymie, Pharmacie et Métallurgie* (Paris, 1786), Vol. I, pp. 447–51. Translated by P. Boucher and C. S. Smith.

Editor's Note

How much of our knowledge of eighteenth-century technology derives from French authors! It was a remarkable period, the "Age of Enlightenment," which drove intellectuals to be interested in technology at a time when technology was neither too complicated nor too extensive to be adequately recorded. (It is well, however, to remember that the great industrial developments in iron and steel production — the coke-using blast furnace, the puddling furnace, and the rolling mill for iron — all occurred in England, without benefit of philosophy or science, and where nobody bothered to write about them.) The great Encyclopédie *edited by Denis Diderot had many emulators, perhaps the most important being the* Encyclopédie Méthodique *issued in many sections by the editor-publisher, Charles Joseph Panckoucke, in Paris, beginning in 1782 though not completed until 1832. Articles were grouped by subject matter, one of the earlier and most important units being the two volumes comprising chemistry, pharmacy, and metallurgy. The writer of the chemical articles, including the one on steel from which we quote below, was Guyton de Morveau (1737–1816). He made a name for himself in both pure and applied science and also as a politician, for he was a deputy to the National Convention that in 1792–93 declared France a republic and voted for the execution of Louis XVI; he later served as secretary and president of the Committee on General Defense and its notorious successor, the Committee of Public Safety, which did much to bring science to the service of the new state.*

Though Guyton de Morveau was educated to be a lawyer, he soon turned to chemistry. His first scientific publication was in 1768, on the role of air in combustion. He later made important observations on the crystallization of iron and other metals. A resident of Dijon, he gave a popular series of lectures on chemistry at the academy of that provincial city and published them there in three volumes in 1777–78.

Practical chemistry interested him as much as theory, and he made notable contributions to fumigation, to the manufacture of soda and glass, and to the development of the hydrogen balloon. He made quantitative studies on the reaction between acids and bases that influenced the later theories of combining proportions, and he improved the Wedgwood pyrometer. He was an early convert to the anti-phlogiston chemistry of Lavoisier, with whom, as with most of the leading chemists at the time, including Priestley and Bergman, he corresponded. He is best known for his work on the reform of chemical nomenclatnre, which he began in 1782. In the extremely influential Méthode de Nomenclature Chimique, proposée par MM. de Morveau, Lavoisier, Bertholet [sic], & de Fourcroy (Paris, 1787), it will be noticed that his name leads that of the other famous chemists. Altogether it was natural that he should be asked to contribute to the new Encyclopédie Méthodique.

Guyton de Morveau was asked to undertake writing for the Encyclopédie in 1780. The alphabet caused him to begin the article on steel (acier) soon after this, and when he had made a critical examination of the phenomena and of the existing chemical theories, he soon saw that Bergman's phlogiston theory was too complex, and that carbon alone, as a material substance alloyed with iron, mainly accounted for its properties.

The entire article on steel occupies pages 420–51 of Volume I of the chemical volumes of the Encyclopédie Méthodique. The fraction of it that we reproduce was preceded by a discussion of the history of the knowledge and understanding of steel, appreciative summaries of Bergman's De Analysi Ferri (based on Grignon's translation) and Rinman's Forsock til jarnets historia, a report of a *number of new experiments by Guyton de Morveau himself, and a section in which the roles of the calorific principle, phlogiston, and the newly discovered vital air are discussed and separately dismissed as bases for the distinction between iron and steel. This article was written before the author knew of the work of Vandermonde, Berthollet, and Monge (see Chapter Nine), read before the Academy of Sciences in May 1786, and was already in type at least by September 1786. Publication of both papers was slow — the volume of* Mémoires *containing the academicians' work did not appear until 1788, and Guyton de Morveau's first* Encyclopédie *volume emerged in 1789. Advance notice of both, however, appeared in 1786.*[a]

There are many similarities between the two treatments, both in the historical part and in the statement of the new ideas. Neither has any use for phlogiston. Guyton de Morveau is more meticulous in his survey of prior knowledge than are the Academicians, has more detail on the chemistry of the "plumbago" extracted from steel and cast iron, and has less original experimental data. Both papers depend heavily upon Bergman, and both conclude that the presence of charcoal or plumbago is responsible, simply, for the difference between steel and iron.

[a] *See Guyton de Morveau,* Observations sur la Physique, *1786, 29, 132–200 and 308–12. The biographical material in this note is based on J. R. Partington,* A History of Chemistry, *Vol. III (London, 1962), 516–34, and the biographical article by W. A. Smeaton in* Ambix, *1957/8, 6, 18–34.*

On the Nature of Steel and Its Proximate Principles

IF, then [*as the author has shown in the preceding section of this article*], caloric matter, phlogiston, or vital air (either alone or when supposedly combined in different proportions) are not enough to provide a satisfactory theory of the conversion of iron into steel, there must necessarily be some other principle that changes and modifies one or the other of these substances by virtue of its particular properties; the beautiful analysis of the illustrious Bergman indicates that this is *plumbago*.

If cast iron, steel, and wrought iron are dissolved in pure vitriolic acid diluted with two parts of water, with the aid of a little heat, the wrought iron will be completely dissolved, or at least there will remain only very little insoluble matter; a black powder will separate from the cast iron and from the steel, in different proportions but always in very perceptible quantity and always more abundantly from the cast iron than from the steel. In the public courses of the Dijon Academy, in order to demonstrate in an unequivocal manner the phenomenon offered by this dissolution, I ordinarily use a gray cast iron that is soft to the file and place in a flask a single thin sheet of it, weighing approximately $\frac{1}{4}$ to $\frac{3}{8}$ ounce and having all of its surfaces filed until they are like new. If the solution is carried out under the very gentle heat of a lamp, a great number of very thin black flakes are soon seen floating in the liquid. I decant the solution before it becomes saturated, pour pure water several times over the undissolved matter, and finally simply let it drain on filter paper.

The following day the piece of cast iron is found to have the same shape [*as originally*] and appears to have hardly decreased in volume, although it has often lost more than one-third of its weight — the precise loss depends upon the moment at which the dissolution is stopped. It is already covered with rust, and beneath this rust lies black powder still adhering to the cast iron, though it is very easily detached. When all of this crust is removed, it ordinarily weighs 12 to 15 grains. The diminution of the volume of cast iron in proportion to the [*extent of*] dissolution is then very perceptible.

On the other hand, the flakes that I have mentioned are recovered on the paper. They are sometimes fully as long as the piece of which they formed a part, and they often recall the shape of it from two sides at once, just like portions of a sheath. At their inner surface they present striations or crosshatchings (not deep but very distinct), and they completely preserve their original black color.

In this very simple operation, it becomes very easy to grasp the important differences that characterize the black powder that adheres to the metal and the flakes that have separated from it. The first is a true ethiops that is very sensitive to the magnet, dissolves in acids, and rusts in less than twelve hours if it is left in open air before being dried. The flakes have none of these properties, any more than does the black powder that is merely the detritus and is also separated from the metal. A magnet that lifts the piece of cast iron even before it is stripped of its crust cannot displace the smallest of these flakes weighing $\frac{1}{32}$ of a grain. They are insoluble even in muriatic acid. When crushed in a piece of agate using an agate pestle, they coat these with a completely bright micaceous skin, resembling a [*black-lead*] pencil.

There is therefore a substance in cast iron and steel that is not given in the analysis of wrought iron, at least in comparable amounts, and that is neither ethiops nor saffron of mars.[b] The

---

[b] *I.e., neither ferric oxide nor ferric hydroxide in today's chemical terminology. The residue of "éthiops" in the preceding paragraph would actually have been iron carbide, not the oxide.*

experiments of the celebrated Scheele have made known to us this singular substance that can be entirely resolved into gas, as though it had no elements that are solid or capable by themselves of acquiring concrete form. (See article "Plumbago" [*in the Encyclopédie Méthodique*].) But even if we had doubts as to its true composition and though we might eventually come to discover some other grosser principle, or even a little metallic earth in a state that we have not as yet suspected, it would be no less certain that this compound as such makes up an essential constituent part of cast iron and steel, and likewise that we do not find it in iron. It is easy now to show the immediate cause of those stains that acids consistently produce on cast iron and steel and that are not observed when soft iron is touched with the same acids: they are due to the precipitation of plumbago.

To these proofs by analysis, we can add a great number of proofs by synthesis: the first and most direct is the production of steel by the mixture of cast iron and wrought iron. Since one is charged to excess with what the other lacks, the sharing that occurs [*on mixing*] can give the mean composition that constitutes steel without any other condition. This is what we have seen happen. (Experiments on iron, Nos. 32, 33, and 34.)[c]

If cast iron is a kind of steel surcharged with plumbago, then steel when melted with plumbago or with charcoal dust should recover the characteristics of cast iron; the event confirmed this analogy. (Experiments on steel, Nos. 13 and 16.)

For the contrary reason, cast iron when treated with the same substances necessarily departs proportionately farther from the state of steel; three decisive results have verified this consequence. (Experiments on cast iron, Nos. 8, 17, and 21.)

With the exception of a few rare cases, the anomaly of which is not even difficult to explain, all the facts that I collected rally naturally to support these principles. We have seen that fluxes or

---

[c] *These experiments were reported in the first section of this article (pp. 430–31), not here translated.*

cements of opposite natures were necessary to operate on cast iron and on wrought iron. We can easily follow this opposition in all procedures that are not complicated enough [448] to conceal the trace of the direct means, for it is the barren cements that act consistently upon cast iron and the carbonaceous ones that determine the conversion of wrought iron [*into steel*]. The former cements soften cast iron, the latter render wrought iron more fusible and more susceptible to hardening on quenching. A bar of cast iron exposed to the same fire, surrounded, in the same crucible, half by bone ash and half by a mixture of bone ash and powdered charcoal, is found after the operation to be half ductile steel, half cast iron as before (No. 21). Can one refuse to say with the celebrated Rinman that in the mixture one of the ingredients truly prevented the effect of the other?

In order to put the truth of this explanation in a still more striking light, I shall pause to consider for a moment the characteristics that associate or, to express it better, that identify, charcoal and plumbago to the phenomenon with which we are concerned.

If the first steel had been prepared by the cementation of iron in this non-ignitable, micaceous substance that we call plumbago, it would not have taken long to suspect that iron retained a part of it; and if the increase in weight of the iron during this operation had been observed, it would have been thought that no other proof was needed of the possibility and the actuality of this union. But only charcoal was used as a cement, just as it alone is still used today. No one even remotely imagined that it [*charcoal*] could combine with metals or even that it could furnish metals with anything other than phlogiston or the metallizing principle. It is this, without doubt, that has kept us from recognizing heretofore that iron acquired the properties of steel only with the addition of an appropriate quantity of this matter.

We know now that plumbago makes niter detonate in the same way that charcoal does and that likewise it leaves the alkali of the niter in an effervescent state; that is to say that it is decomposed and is resolved into inflammable gas and mephitic gas.[d] The observations that I published in the *Mémoires de L'Académie de Dijon* (1783, ser. 1, p. 76) prove again that there are carbonaceous substances that have all the external appearances of plumbago and yet which are neither combustible nor ignitable like charcoal. I shall collect elsewhere many other analogous facts. (See "Charcoal" and "Plumbago" [*other articles in the Encyclopédie Méthodique*].) There is one, however, that seems to belong here.

Last October, with M. [*Smithson*] Tennant, I performed some tests with the blowpipe; tired of finding only charcoals that crackled as soon as we put the point of the flame to them, we thought of exposing them for an hour to a great fire, in well-sealed crucibles, to remove the humidity that seemed to us to be the cause of this decrepitation. However, we were soon convinced that desiccation was not the only effect of the operation which we had them undergo. They refused after that to become inflamed and grew red-hot without being consumed: they had therefore perceptibly approached the condition of plumbago. From that we can conclude that the charcoal that has been used in an earlier cementation is no longer exactly the same as it was before, and if (as all Metallurgists assert) it is equally suitable for converting new iron into steel, it is a proof that it acts really only as plumbago. We would find many more proofs of this by scanning the experiments on cementation done with animal carbonaceous substances that refuse to ignite and which consequently have a much more nearly perfect resemblance to plumbago.

When for the first time I saw real plumbago separate from M. d'Arcet's alloy[e] by fusion in boiling water — an alloy that

---

[d] *The effervescent alkali remaining after the burning of graphite or charcoal with niter is potassium carbonate, which evolves CO_2 when acid is added. The "inflammable and mephitic gases" are respectively carbon monoxide and carbon dioxide, although the former may have been confused with hydrogen.*

[e] *D'Arcet's alloy contains 50 per cent bismuth, 25 per cent lead, 25 per cent tin. It melts at about $93°C$. The modern metallurgist will share Guyton De Morveau's astonishment that graphite floats out of it when it is melted!*

I had prepared myself with very pure and well-melted metals — I was loath to believe that it had really been alloyed with these metals; but in comparing this to what I said in the same *Mémoire* regarding plumbago from furnaces in which iron ores are melted, regarding procedures that gave me a true *Eisenman* or artificial micaceous iron ore, regarding the experiments of Bergman and of Rinman, and even the highly striking analysis of gray cast iron by weak vitriolic acid that I reported earlier, it appears impossible to doubt that metals can combine with plumbago, even when in the state of a regulus, and especially iron. It will be a new exception to the overgeneralized rule that had already led us so much in error, that metals could combine only with metals. Since plumbago is a mephitic sulphur, this phenomenon should not be more astonishing than any other pyritic combination. (See *Alloy*.)

I shall not dissemble the fact that M. Grignon, in the notes to [*his edition of*] M. Bergman's *Analyse de Fer*, strongly objects to this combination of plumbago with iron; that he maintains that it is heat alone as it accumulates that effects the conversion; that he finds support for this in that the same charcoal may serve for several cementations; that he thinks that plumbago, like calcareous earth, is rather disposed to absorb the steelifying principle [*principe aciérant*] — in a word, that the black powder that precipitates from steel during its dissolution by vitriolic acid is probably only an interposed separate body — that known as cinder [*i.e., slag*] — that does not make up an integrant and constitutive part of it (pp. 69, 104, 154). However, after weighing all these objections, he [*Bergman*] wrote to me on the 18th of November 1783 saying that he believed these doubts to have been ill founded. I shall give here in his own phraseology the passage from his letter that seems especially designed to inspire confidence by the testimony of the conviction that he held regarding the exactness of his experiments.

The great question between us is in regard to plumbago, which M. Grignon affirms is accidental in pig iron and steel. I shall honestly acknowledge my error as soon as he is able to send me a single piece of pig iron and of steel that is free from it. I pray that you also search [449] diligently for samples and send me some, for if I am wrong, I wish candidly, my dear friend, to be undeceived as soon as possible. M. Grignon is perhaps prejudiced in favor of his system, and thus I do not find it astonishing that my plumbago has saddened him; but if the steelifying principle consists of phlogiston and caloric matter, it seems to me that plumbago, which contains much of them and is more corporeal, should not be regarded as incompatible with this system.

I do not see in fact what one can bring in opposition to such repeated analytical experiments, unless it be an even more exact analysis that would contradict the original results or at least that would change the conclusions through a more precise determination of the nature of the products. That is what M. Grignon did not undertake, and he even confirms by his own observations what I had said regarding the plumbago that is frequently found in the slags of forges: *cast iron surcharged with phlogiston* (these are his terms) *produces a great deal of a substance that is separated by the action of fire alone, and that resembles plumbago in many respects...It is scaly, black, silken and light. It stains the fingers when crushing it, and has the slipperiness of plumbago.* We can henceforth consider it as well demonstrated that cast iron and steel really do contain a perceptible amount of a substance that is not iron in the metallic state and that has the property of remaining united with iron.[1]

---

[1] This article was in the press when, through the courtesy of my illustrious colleague M. [Richard] Kirwan, I received the latest work of M. Priestley (*Experiments and Observations* [*relative to the various branches of Natural Philosophy*...], Vol. III, Birmingham, 1786). By what I shall extract from Section XXIV, the reader will judge the satisfaction I had of finding therein some of my experiments and of being able to offer still another guarantee of the truth of those facts that serve as the basis for the theory that I have adopted.

A comparative examination was made before and after cementation of the cast iron nails or bolts that in Birmingham are cemented in charcoal in order to give them a sort of ductility and are therefore called nails of annealed cast iron [*fer crud engraissé, i.e., malleable cast iron*]. This great physicist observed that 1000 grains of this cast iron, perfected in this way, left 68.75 grains of insoluble black powder after their dissolution in dilute vitriolic acid; this dissolution occurred very slowly,

But how does plumbago act in the conversion of iron into steel? How do plumbago and charcoal-turned-plumbago penetrate the entire mass of the bars of iron during cementation? Finally, how can such a small quantity of plumbago produce so great a difference between iron and steel? I admit that these questions are not yet without difficulties, and the last especially appears to M. Bergman to be the Gordian knot of the theory of this operation. However, if we attentively consider all the circumstances of these phenomena and carefully bring together the facts that necessitate this conclusion and those that most oppose it, we soon become

---

black flakes separated from the piece, but its shape was preserved; that steel in general gives much more black residue than iron does, which could cause it to be called annealed malleable iron; that this black residue is not soluble in muriatic acid;[f] that if 10 English grains of this residue (8.119 French grains) are exposed to the focus of a lens [*in vacuuo*] they furnish $1\frac{11}{18}$ French cubic inches of mephetic gas and 12.889 of detonating inflammable gas, although a large part of this light powdery substance is dissipated; that this residue was reduced to $\frac{6}{19}$ of its weight when it was melted by the lens in open air, and it then resembled slag. In a word, that it behaves like plumbago and, like it, is resolved almost entirely into mephitic acid gas and inflammable gas.

By taking the mean term of his [*Priestley's*] experiments, it can be seen that 98.38 French grains of crude iron, before cementation, give 145 French cubic inches of inflammable gas; that a like quantity of cemented crude iron gives $169\frac{1}{4}$ cubic inches; that a like quantity of steel gives 155 cubic inches; and that 97.15 grains of iron, from which one of the steels had been prepared, furnished $155\frac{1}{4}$.

It is the opinion of those who manufacture steel (also says M. Priestley) that iron neither decreases nor increases in weight during the operation [*of cementation*], and those who cement cast iron [*i.e., make malleable cast iron by annealing*] affirm that it loses a lot; but his own observations agree more closely with those that I reported on the subject: he saw 72 grains of iron acquire an increase of 3 grains on cementation; it was 6 in 1440 [*grains*] of cast iron.

Some of the results of M. Priestley's experiments seem to indicate that the crude iron that is subjected to cementation in Birmingham is of the nature of those cast irons that are called white, or at least light gray; for the dark gray cast irons alone have the property of being filed, drilled, and even of being compressed to a certain degree. They also yield a much greater quantity of plumbago. They would not be capable therefore of either the same amelioration or of an equal increase in weight.

[f] *Priestley actually said that* "As a greater quantity of inflammable air was procured from annealed cast iron than from malleable iron, I think it may be concluded that steel (which is malleable iron annealed) will always give more

aware that the latter surprise us only because they are outside the class of things that are most familiar to us: this assuredly is not a reason for denying the possibility.

There are several ways of thinking of the action of plumbago. Independently of the change that it brings to the composition when by itself, it can (like its congener charcoal) appropriate a portion of the vital air that remained joined to some portion of the martial earth and thus render the metallization more perfect and more uniform throughout the mass. On the other hand, M. Lowitz discovered that carbonaceous substances had the property of removing, even in the humid way, the empyreuma that we can attribute to hardly anything other than the presence of a fixed inflammable principle.[g] If it is true that iron must lose some phlogiston in order to become steel, as Bergman thinks, and as is indicated by the amounts of inflammable gas that he obtained from the one and the other, would it not be possible that charcoal with the aid of heat could also produce [450] an analogous dephlogistication? It is precisely because we still have only very little light on this

---

air than iron, provided that, in the solution, it should yield no greater a quantity of black residuum. But in general there is much more of it found in the solution of steel than in that of iron." *Moreover he remarked that the residue from steel was partly soluble in muriatic acid, as, indeed, iron carbide would be.*

The reference is to the second set of Priestley's essays, sometimes regarded as Volumes IV to VI continuing the earlier Experiments and Observations on Different Kinds of Air *(London, 1774, 1775, and 1777). A French translation of the volume in question appeared in 1787. Guyton de Morveau's conversion of the weights and measures into French grains and cubic inches are not all correct. Priestley remarks on the variability of his results, and he attributes some of the gain in weight of the cast iron on annealing to the formation of scale. His "annealed" or "cemented" cast iron, of course, is whitehart malleable, though Priestley seems to be under the impression that the Birmingham foundry that supplied it to him had annealed it packed in charcoal. Altogether, Priestley's work on iron is not on a high level and adds little to what was already known on the basis of Bergman's experiments.*

The modern reader should be alert to the distinction between malleable iron, i.e., wrought iron, and malleable cast iron, which is annealed white cast iron.

[g] *J. T. Lowitz had just published a paper in Crell's* Annalen, *1786, 1, 293, reporting the discovery of adsorption on wood charcoal. He used discolored solutions of tartaric acid.*

singular property of charcoal that we are permitted to form conjectures in the search for its effects.

Since bars of iron placed in the chest of the cementation furnace have been considerably rarefied and even softened by the violence of the heat, it is not at all disagreeable to imagine that in the long run plumbago can penetrate to their interior (all the more so since this substance is itself placed by the fire into a nearly aeriform state), that there is some sort of affinity between these two bodies that determines their union, and that it is an effect of this affinity to favor the transmission of the fluid body throughout the solid body, as we shall see elsewhere. Here we have proofs that the conversion is effected only successively from the surface to the center, for if the fire was not kept up long enough, the core remains iron; this slow progression alone would be sufficient to characterize a substance different from the calorific principle, the communication of which is much more rapid.

The observation of the celebrated Rinman that iron becomes steel without coming into immediate contact with any carbonaceous cement, provided that the crucible is surrounded by charcoal and placed in the cementation box, might make one think that this conversion depends in fact upon a substance more subtle than plumbago; but first, it cannot be the calorific principle, since this metallurgist assured himself that conversion does not occur in a glass cylinder, even though placed in the same cementation chest. (See the experiments on iron, Nos. 6 and 7.) It would be more reasonable, therefore, to conjecture that charcoal, or plumbago, furnish to the new composition only their gaseous elements, which are more capable of penetrating the crucibles and the inactive cements, and thus that a portion of the iron itself combining with these gases, or only with the mephitic gas, regenerates the plumbago right in the interior of the masses: but this supposition appears useless to me once it has been verified that plumbago itself can be put by fire into a vaporous state. Furthermore, it seems to me that without making any hypothesis the increase in weight of the steel and the products of its analysis form a strong enough and more direct proof of the composition of iron in its passage to the state of steel.

When there is a new composition, one should no longer be astonished that there are different properties. It is a necessary consequence not only of the change in the proportions of the substances that exercise simple affinities but also of the affinities that are acquired by the product of their union. All that we can desire to add to the probability that these property changes [are caused] by the sole effect of such a weak accession of matter is that the same combination always brings about analogous modifications that are, so to say, proportionate to the amounts. That is what we find here, by comparing the degrees of fusibility of cast iron, steel, and iron; their disposition to rust; the intensity of the stains left by acids; and by considering that steel becomes untractable like cast iron in plumbago; that gray cast iron hardens perceptibly on quenching, etc., etc. After this to ask why iron, turned to steel, no longer entirely resembles iron is to ask why copper does not absolutely resemble brass.

Let us conclude therefore that steel, in whatever manner it may have been formed, is nothing but iron that approaches the nature of ductile iron, because the martial earth in it is freer from heterogeneous parts and, if not more perfectly metallized, is at least more completely so than in cast iron; that it differs from iron because it takes into its composition a perceptible quantity of plumbago; that it approaches cast iron even more than ductile iron because of the presence of this mephitic sulphur; that it differs from gray cast iron only in that this sulphur is much more abundant in the latter; that it is farther removed from white cast iron because the latter conceals earthy, nonmetallized, or even foreign parts that can be separated from it by a second tranquil fusion in closed vessels and without addition; that in all cases, the transition of cast iron to the state of steel is thus done by purifying the iron and removing the excess plumbago; that the conversion of iron into steel is accomplished principally because a perceptible quantity of plumbago is formed there or is received into it; that first of all heat exerts influence in these changes only by producing and maintaining fluidity, without which no combinations are made; that the composition that constitutes steel can very well fix, by its own affinity, a greater quantity of caloric matter; in a word, that the general properties of steel depend upon a precise addition of these principles, just as the different qualities of steel depend upon the accidents that vary the proportions.

It will perhaps be thought that I could have dispensed with reporting such a large number of experiments and observations in order to establish this conclusion; but it concerned one of the most obscure, most controversial, and at the same time most important points of the theory of the chemistry of fire and of metals. I could make it clear only by a comparison of the facts; before comparing them, it was necessary to establish them; since several are as yet only written in foreign works, it was necessary to show them in sufficient detail to make them understood; in any case, who does not know that facts constitute the most useful part of a work of science? We like to find them there, even though new discoveries make us feel the necessity of embodying them in another hypothesis.

When these principles are better known and put to the test by the friction of opinions and begin to obtain general assent (which, for most people, is the best of all demonstrations), they will probably serve to perfect the manufacture of steel; but until then one should guard against too lightly abandoning the procedures that are in use and even the manner of studying them, which has appeared to assure the success of these large factories, all the more so because the best contrived innovation often serves only to occasion considerable losses; therefore it will not be astonishing that the author of metallurgy guided his work according to this scheme.[h]

[h] *The separate article on steel metallurgy that follows the present in the* Encyclopédie Méthodique *after a brief description of compounds made for pharmaceutical*

In chemistry, we must henceforth, with the illustrious Bergman, consider it as unquestionable that whenever the properties of iron are being worked on, it is not steel, as we believed, that should be taken as a subject for experiment but ductile iron, which is the purest metal of this sort and, if we may say so, the most iron.

uses from steel (as the supposed purest form of iron) was written by J. P. F. G. Duhamel, professor of metallurgy at the Ecole des Mines, and inspector of mines. It is a good practical report on steelmaking by various methods, with especially good detail on cementation and a description of English cast steel, both in the main following Gabriel Jars. Though in general avoiding theory, at one point Duhamel remarks that in order to make good steel it is necessary "to remove heterogeneous substances and to communicate phlogiston to [iron], or at least to conserve that with which it had been charged in the earlier operations."

CHAPTER TEN

On the Different Metallic States of Iron

Charles Auguste Vandermonde, Claude Louis Berthollet, and Gaspard Monge

"Mémoire sur le fer consideré dans ses différens états métalliques," from the *Mémoires de l'Académie Royale des Sciences*, 1786, pp. 132–200. Translated by Anne S. Denman and C. S. Smith.

Editor's Note

If one could read only one paper in the history of iron and steel before 1800, the selection would probably be the following one by C. A. Vandermonde, C. L. Berthollet, and G. Monge, which summarizes the old viewpoints and procedures and clearly states the new theory that was to inspire research and improvements in production through most of the nineteenth century. It was read before the Academy of Sciences, Paris, in May 1786, though the volume of Memoirs for that year was not published until 1788.[a]

French chemists at the time were, under the leadership of Antoine Lavoisier, making many significant discoveries and developing the new framework of chemical theory and associated nomenclature that permitted many long-known chemical phenomena to be more simply understood. The old terms "element" and "compound" took on a new and more definite meaning. The studies of the compounds of the newly discovered oxygen with other elements made it obvious that the major differences between calces and metals lay in composition, not in the evanescent principle of phlogiston that had occupied chemists for a century. It was natural enough, therefore, that the French chemists should seek in simple compositional terms the explanation for the manifestly different properties of the different forms of iron.

There has never been a time when knowledge of the practical metallurgist interacted more closely with the very forefront of advancing

[a] *The paper was reprinted in* Observations sur la Physique, *1786, 29, 210–87. A German translation appeared in 1794 — "Über das Eisen in seinem verschiedenen metallischen Zustande,"* Chemische Annalen (Crell), *1794 (i), 353–83; 460–73; and 509–28.*

scientific theory: first in the unraveling of the role of the atmosphere in oxidation and reduction (for it was the old smelting and assaying operations that exicted Lavoisier's mind) and next in the question of the nature of steel and its distinction from iron.

Torbern Bergman had sent a copy of his 1781 paper on the analysis of iron to the Academy of Sciences in Paris, where it was received by the chemist Berthollet and eventually transmitted to Grignon, who published a translation of it in 1783. This aroused the interest of Guyton de Morveau, then at work on his encyclopedia article in the provincial city of Dijon (see Chapter Nine in this book), and precipitated the collaboration of three other scientists of the very highest rank in the writing of the present paper. It is a chemist's paper, and it is indicative of the character of science at the time that only one of these — Berthollet (1748–1822) — is primarily identified as a chemist. His work had a strong industrial flavor. He published papers on both bleaching and dyeing that came from his period as director of the Gobelins tapestry works, but historians know him best for his studies of chemical affinity, his approach to the law of mass action, and for his opposition to the law of constant combining proportions which played such an important role in the development of chemical theory early in the nineteenth century.

C. A. Vandermonde (1735–1796) is primarily known for his contributions to the theory of equations and other branches of pure mathematics, but he was closely associated with the leading chemists, and it was in the institution of which he was director — the Conservatoire des Arts et Métiers, *founded by the* mécanicien *Vaucanson — that the experiments on iron were carried out.*

Gaspard Monge (1746–1818) was a mathematician of the highest rank who founded descriptive geometry and was later important in French politics, especially under Napoleon. He had been an instructor of mathematics at the celebrated military school at Mézières and had a strong interest in useful technology. His development of projective geometry (kept as an official military secret for several years) came about from his annoyance at the primitive computational techniques used by stone masons employed on military structures. He exploited the geological and industrial richness of the region around Mézières by taking his students on frequent field trips to local plants and workshops. It was probably this that started his interest in steel, though he must have had many conversations with the metallurgist Clouet, who was one of his colleagues at Mézières and later published papers on methods of making crucible steel and on the production of textured steel of the Damascus type. But Monge had also worked in pure chemistry and had attracted the attention of chemists by the work that he did at Mézières in the Summer of 1783 on the quantitative synthesis of water from the gases oxygen and hydrogen. This is one of the great discoveries of that great decade. (Cavendish in England and Lavoisier in Paris independently did similar experiments; the former actually preceded Monge, though he did not publish until later.) Monge later moved to Paris (continuing to commute to Mézières for classes for some years) and later became the leading scientist-politician of France, to be eclipsed only after the fall of Napoleon.

The background of the association of these three men on this topic would be an interesting study. Their close association continued for many years, and their names in various combinations figure largely in any discussion of the role of French intellectuals in the Revolutionary and Napoleonic periods. In 1796, Monge and Berthollet were sent with a committee of artists to collect the aesthetic plunder demanded from Italy by Napoleon, and they jointly accompanied Napoleon on his Egyptian campaign in 1798. Monge was the first president of the Institut Egyptien. All three of the men had a strong influence in the establishment of technical education in France. Monge and Berthollet founded, in 1801, the Société d'Encouragement pour l'Industrie Nationale which, by its prizes, its meetings, and its publications, greatly stimulated the industrial arts and the related sciences.

All three played lively parts in the production of armaments for the Revolution, in an atmosphere not unlike the mobilization of scientists in World War II. They ran intensive courses on the production of saltpeter, gunpowder, and cannon, and supplemented this by writing instructional pamphlets. On 18 Pluvoise, An II (7 February 1794), Monge was ordered by the Committee of Public Safety to write on gunfounding and produced a book of 231 pages, Description de l'Art de Fabriquer les Canons, *which was promptly issued by the printing office of the Committee. Among its sixty instructive plates are three showing how to convert chapels and churches into gun foundries for the public good, and there are full details of a short-cut method of making cannon molds by the use of stacked flasks of dried sand as well as the traditional loam molds made more laboriously according to the ancient bellfounders' practice. Vandermonde had previously been impressed by the Committee to write a description of the methods of making white arms (swords and bayonets), based on a hasty study of the Klingenthal works near Strasbourg. Vandermonde, Monge, and Berthollet were associated again in writing a 44-page pamphlet,* Avis aux Ouvriers en Fer sur la Fabrication de l'Acier, *which was published by order of the Committee of Public Safety in December 1793. Thirteen thousand copies of this were printed, and it was ordered that it be distributed* "avec profusion." *Dissemination was helped by immediate reprinting in* Observations sur la Physique, *1793, and it was also included in Volume 19 of* Annales de Chimie, *the first to appear, in 1797, after the hiatus caused by the Revolution. It begins with a tirade against the tyrants of England and Germany who had interrupted commerce in steel and an exhortation to French workers to make their own steel for the arms needed by the citizens to insure their freedom from slavery, then goes on to an excellent brief account of the nature of steel and its manufacture, all in terms of the new theory.*

The order of the authors' names on this publication is interesting: Berthollet was perhaps lacking in revolutionary enthusiasm, but he clearly had a hand in the writing. He had previously given, under his sole authorship, a popular form of the new theory under the title, Le

Précis d'une théorie sur la Nature de l'Acier, sur sa préparation et ses différens espèces (Paris, 1789). *Appended to this is a description of work done by Chalup and Clouet at Mézières on the use of a glass flux to prevent decarburization during the melting of crucible steel.* Berthollet says that he wrote the Précis *specifically at the request of the Minister of Commerce, who wanted France to be free from the need to import steel. Many paragraphs from Berthollet's* Précis *are included in the revolutionary pamphlet, which also quotes at some length the description of the English manufacture of cementation steel given in Gabriel Jars'* Voyages Métallurgiques, *Vol. I, Lyons, 1774, and reproduces his plates showing the furnaces.*

The Academy paper that first resulted from the collaboration of these three men is a masterpiece of clear thinking. It gives a fine summary of contemporary practice and of the only two previous authors whom they deemed worthy of notice, Réaumur and Bergman. The paper was clearly inspired by Bergman, and it leans heavily upon his factual information. The authors give, however, new and more accurate determinations of the gain in weight accompanying cementation and of the volumes of "inflammable air" evolved when different kinds of iron are dissolved in acid. They then discuss the old processes of the making of iron and steel in terms of their new economical and elegant theory, which is devoid of all concern with the phlogiston and caloric in various forms that had made Bergman hard to follow. Their theory is nothing more than that cemented steel is just well-reduced iron combined with a certain proportion of charcoal, that over-cemented steel has more than the right proportion, and that cast iron contains still more charcoal, though it may also, especially when white, retain some of the much-talked-of new element oxygen; that is, white cast iron is iron that is incompletely reduced. This interpretation of white cast iron is, of course, quite wrong, though it is plausible, for white cast iron arises whenever there is a deficiency of charcoal in the blast furnace charge. Bergman had actually given the correct explanation, namely that white cast iron is iron containing more carbon than steel and that gray cast iron also contains silicon, which was later seen to promote graphitization. Neither form of cast iron contains any oxygen. But it is common for the proponents of any successful new theory to overextend it, and the French chemists in their enthusiasm for the newly discovered element and its role in combustion and calcination rather naturally attributed to oxygen other effects in which it was not involved. Their most notable error was to regard oxygen as a necessary constituent in all acids. It will be noticed that, despite their disbelief in phlogiston, the authors still use the term dephlogisticated air for oxygen, for the new nomenclature had not yet been promulgated.

The new carbon theory was rapidly disseminated and had champions and opponents throughout most of Europe. One of its principal champions was the German W. A. Tiemann in his Systematische Eisenkunde mit Anwendung der neuern chemische Theorie *(Nuremberg, 1801). It remained to unravel the various effects of nitrogen as well as the elements silicon, sulphur, phosphorus, and manganese that Bergman had discovered in iron, to demolish the theory of the role of oxygen in cast iron which the French chemists had erroneously promulgated, and above all to elucidate the actual form in which the carbon was present in its various manifestations in iron. Despite many studies of acid-extraction residues, especially those of C. J. B. Karsten, the most important carbide was not definitely identified as Fe_3C until 1881. The different solubility of carbon in the different allotropic forms of iron — the whole basis of the hardening of steel — was first disclosed by studies of the behavior of quenched and annealed steels in acids, then microscopically observed by H. C. Sorby in 1864, to become a central phenomenon in the structural arguments between the leading metallographers in the last two decades of the nineteenth century — all much beyond the time covered in the present book and belonging to a chapter of metallurgical history in which structure replaced composition as the focus of research and explanation.*

The importance of the compositional approach to steel was firmly established during the first half of the nineteenth century and thereafter, and chemical analysts were associated in one capacity or another with all of the great steel works. It was, in fact, through chemistry that science first came to the steel industry, though it was through physics that metallurgy as a whole reached real maturity, for the physical approach was needed to emphasize structure. The future will develop even wider interrelationships among different viewpoints.

The clarity of the distinction between the various forms of iron in the new theory of Vandermonde, Berthollet, and Monge arose from the fact that it dealt with composition alone. It conformed to the latest chemical viewpoint, it was part of the so-called "Chemical Revolution," and it rode on the crest of the new wave that was to dominate science for more than a century and lead to great achievements in understanding. But it was an overemphasis on only part of the whole science of steel. We can now see that it involved a certain blindness, that a real price was paid for the indubitable gains. In explaining the difference between gray and white cast iron, they did not even notice that the main distinction was a structural one, namely whether the carbon was present in the elemental state of graphite or whether it occurred in crystals of iron carbide. The new theory was a great advance over that of Réaumur in chemical terms, but the appearance, the texture, and the "feel" of steel were lost. Its physical properties were regarded by Vandermonde, Berthollet, and Monge as dependent on composition alone. Gone were all the qualities of Aristotle, gone were the sulphurs, phlogiston, and caloric that the eighteenth-century chemists used in their attempts to understand the qualitative changes in the nature that accompanied the different forms of iron. Gone were Réaumur's and Grignon's concern with the shape and arrangement of the parts. In the whole long paper on steel, virtually nothing was said about the mechanism of hardening! Structure came back into science only slowly in the nineteenth century, and the scales of structure of most interest to metallurgists were the last to become part of rigorous science. At first the obvious allotropy of the elements carbon and sulphur was noticed, then, as organic chemistry matured, the existence of isomers was explained in terms of the spatial arrangement of the atoms within a molecule. Excellent mathematical frameworks for understanding crystal symmetry were developed, but the molecule dominated virtually all theories of the properties of matter until after the discovery of X-ray diffraction and its application in the second decade of the twentieth century. Eventually, however (and in large measure beginning under the stimulus of the studies on the microstructure of steel), the old appreciation that properties depended on structure as well as on composition returned. Yet, though one may lament the disinterest in structure on the part of nineteenth-century chemists, it is hard to see how the new chemistry could have advanced had it continued to be concerned with all of the structural complexities and with the fine points of property variation accompanying mechanical and thermal treatment. It was even necessary for chemists, in consolidating the new atomic theory, to ignore the large number of compounds (sometimes called "Berthollides" after Berthollet who insisted on their real existence) that did not fit simple ratios required by Daltonian atomism. Much of the virility of metallurgy lies in the fact that it has been at the borderline between the advancing pure sciences and hence has kept in touch with the best theoretical approaches but, having practical aims, is continually in contact with the real behavior of matter. The phenomena of steel behavior led to the discovery of its microstructure, and in turn metallography uncovered many facts that incited the beginning of the new science of solids, solid-state physics. All of this is beyond the scope of the present volume.

On the Different Metallic States of Iron

THE properties that iron acquires as a result of the different operations that are performed upon it in factories differ so greatly from each other, the results of the same operations performed on iron in different forges are so unlike one another, and even changes introduced in the operations of a single forge by design or by carelessness make so great a difference in the nature of the results, that it has taken a long time to persuade men that iron could be an invariable metal, like gold. It is only recently that we have been led to conclude (initially by analogy) that the variations in the properties of iron that follow different operations, or the same operations in different forges, come only from foreign substances, metallic or otherwise, with which iron is alloyed. That conclusion has been confirmed, at least in some respects, by later discoveries: for example, M. [*Torbern*] Bergman discovered that cold-short iron differs from ductile iron by [*the presence of*] a white precipitate that is obtained on dissolving the cold-short iron in vitriolic acid. This precipitate never occurs when soft iron is dissolved in that acid. M. Meyer showed that this precipitate, which has been named Siderite, was a phosphoric salt of iron [*sel phosphorique martial*]; and this result has been confirmed at Mézières by MM. Clouet, Dulubre, and Chalup, who removed phosphorus from the precipitate by procedures that have been reported to the Academy. M. Clouet also discovered [133] that arsenic is the substance that gives iron the quality of being brittle when hot; at least he found this semimetal in the black residue that remains at the botton of solutions of one kind of hot-short iron in vitriolic acid. [b]

[b] *Sulphur is, of course, by far the most common cause of hot shortness in iron. Its presence was suggested, on the basis of its smell, by Jousse in 1627 and had been confirmed by Bergman in 1781.*

Great strides could be made in this field if it could be determined with more care than has hitherto been taken just what properties are given to iron by each of the different materials, such as manganese with which it is often united in different proportions. But apart from these researches (which are nevertheless necessary), iron considered in its pure state, or at least free from all foreign metallic substances, is available to craftsmen in four different forms. It is fragile and fusible on leaving the blast furnace; ductile and infusible on leaving the finery; through cementation, it acquires the remarkable property of being able to acquire an extreme hardness by quenching; and, finally, cementation carried on for too long a time makes the iron fusible again and untractable to the hammer. What are the substances to which iron owes its properties in these four different states? That is the question that we are asking ourselves; but before presenting our researches we are going to report certain observations on the manufacture of iron which have been useful to us, some of them being new, and to give a short account of discoveries made by chemists who have preceded us and from which we have profited.

The Smelting of the Ore

The first operation that is performed on iron is the smelting of the ore in blast furnaces and the conversion of it into a regulus. If iron ore were simply the calx of this metal, free from any combinations, it would only be necessary to mix it in the blast furnace with charcoal, which would remove the dephlogisticated air from the ore by combustion and would return the iron with variable exactness to its metallic state. But in iron ore [134] the calx of the metal is almost never by itself; it is almost always united with other earthy materials which are only slightly fusible and which protect it against any external chemical agent. It is therefore necessary to mix the ore with a flux which, causing the ore to melt, uncovers the calx and abandons it, so to speak, to the action of the charcoal.

The product of this fusion falls in the crucible that is immediately beneath the blast from the bellows; the earthy materials form a glass, called slag, which is colored and opaque in varying degrees, which floats, and which is made to flow out from time to time by a higher opening [*in the furnace*]. The partly reduced iron gathers at the bottom, where it forms a liquid bath, kept from combustion by the layer of slag that covers it. Approximately every twelve hours it is run out through a lower opening, in large ingots, or else it is poured into molds, depending upon its destination. This is what is called "cast iron."

In that state the iron is generally brittle, fusible, and already has variable properties, not as a result of the different materials that it may have retained from the gangue, which are not our object of study, but principally as a result of the regimen that has been followed in charging the furnace. It is the custom therefore to distinguish between white cast iron, gray cast iron, and black cast iron, according to the color that the fracture displays.

White cast iron is brilliant in its fracture and is crystallized in large facets; it is harder and more fragile than the others. It is never used for works that must resist a certain stress, and which must then be repaired with tools.

Gray cast iron, whose fracture is matte and grainy, is more flexible than the preceding one and is more easily cut. The artillery pieces of the Navy are required to be made of this material, for they must be pliant enough to resist the violence of the powder, and since they are usually cast solid, [135] they must then be bored on a lathe. This substance is also crystalline, but its crystallization is more confused than that of white cast iron. One of us happened to be in an iron foundry where cannons of gray cast iron were being cast for the service of Holland when a piece of scrap was being sawed in order to put parts of it into the reverberatory furnace. The section passed exactly through a cavity or flat fissure, directly perpendicular to the axis of the piece, which had probably been caused by the shrinking of the metal. The sides of this cavity were uniformly covered with rounded projections which bristled with a regular crystallization.

Finally, the black cast iron is even more rough in its fracture; it is composed of less adherent molecules, which crumble more easily. It has no use other than to be remelted with the white cast iron.

These three states of cast iron have no relation to the qualities of the wrought iron that result from fining. Whatever color the cast iron may be, it is impossible to judge at a glance the nature of the wrought iron that will be obtained from it; and whatever the nature of the ore, it is always possible to give to the cast iron whichever of the three preceding characters is desired. If in the smelting of the ore as little charcoal is used as possible, the cast iron is white; it becomes gray when the amount of charcoal is sufficiently increased in charging the furnace; and finally it becomes black when the use of that combustible is greatly increased. Thus charcoal is the only cause of the color shown by the fracture of cast iron and largely determines its imperfect ductility and the greater or lesser ease with which it is machined.

In some factories it is necessary to remelt cast iron in reverberatory furnaces, either to obtain a greater quantity of material for casting larger pieces or to cast pieces that cannot be produced [136] with sufficient care from the blast furnaces. Every time that originally gray cast iron is remelted in that manner, especially when the fire is made hotter to make the iron more fluid, not only does it become whiter and whiter but it also becomes more like wrought iron; as if the charcoal to which the cast iron initially owed its color, being consumed without contact with air, was partly destroying that which was giving it fusibility.

Iron, as we know, is combustible. It burns in free air and burns even better in dephlogisticated air; but when cast iron is heated to incandescence in air, either in a blacksmith's forge or in the focus of a burning glass, its combustion presents a phenomenon that is not observed with ordinary iron — it shoots off from all parts an endlessly succeeding shower of sparks, which split up in the air when they are already far from the mass. This effect is more pronounced as the cast iron is grayer.

In furnaces where bombs and cannon balls for artillery service are cast and where the iron that is used is midway between white and gray, the metal in the crucible is taken away in great ladles made of wrought iron and coated with clay to prevent their dissolving in the material that they are to contain. These thick masses, always colder than the molten iron, cool the part that they touch and harden the least fusible portions of the metal. When the batch has been poured into the mold, the ladle is always lined on the inside with a quite considerable layer of plumbago [*plombagine*] disposed in little sheets, like mica. The quantity of that substance is always greater when the cast iron is grayer, that is to say, when more charcoal has been put into the furnace charge. As to the nature of this product, it is the same as the graphite in English pencils: it is soft to the touch, it leaves [137] marks on paper, and it resists the action of fire in laboratory blast furnaces when it is shielded from contact with air.

The Fining of Cast Iron

The second operation that is performed on iron is *fining*[c], or its conversion from the state of cast iron to that of wrought iron. For that operation, the cast iron is placed on the hearth of a forge, in the midst of coals enlivened by the blast from two bellows. First it melts down there, then the finer, exposing it continually to the blast, maintains it for several hours at a very high temperature and renews its contacts with the charcoal. It loses its fusibility little by little, it assumes a pasty state, and the worker forms a kind of mass

---

[c] Affinage. *Though it seems mildly absurd to designate nearly complete purification as "fining" and incomplete decarburization as "refining," we have thought it best to conform to eighteenth-century English usage and to use the latter word only when chemical sense demands it. The process referred to, of course, is the finery process that was carried out in a hearth, prior to the introduction of the puddling furnace. For the contemporary refinery process, see fn. h, p.144.*

from it which is called a "bloom" [*loupe*]. This bloom is then put under a large hammer which, by a great compression, presses from it and flings far away all those parts that, still being too fusible, partake too much of the nature of cast iron. What remains on the anvil is elongated by the hammer and, after many repetitions, finally takes the form of a bar which can then enter the market under the name of "wrought iron" or "fined iron."

By the operation that has just been described, the properties of the iron are very much changed; previously it was fragile, it was crushed instead of elongated under the hammer, and it became perfectly liquid in the fire; now it is ductile, it extends under the hammer, can be folded, slit, rolled, and drawn into wire; finally, it is no longer fusible, and the most violent fires from our workshops can only soften it and bring it to a pasty state.[1] However, it still

---

[1] When we say that perfectly refined iron is infusible, we do not mean this rigorously. On the contrary, it is very probable that forged iron, which becomes pasty in the fire of our forges and which softens increasingly as the temperature increases, would finally pass into a true fusion if there were some way to produce and sustain a much higher temperature. We have even had occasion to assure ourselves of this at the Creusot foundry near Montcenis in Burgundy, where furnaces have been recently established, [138] heated in the English manner with mineral coal, and blown by steam engines. The great size of those furnaces, the speed and volume of the blast at the tuyère, and the density of the coal cause the temperature of the hearth to be much higher than that of ordinary furnaces in other forges. (Through the orifice of the tuyère we exposed to that temperature some $\frac{3}{4}$-inch square bars of excellent Franche-Compté iron, and in less than a minute the bars had melted and run into the crucible. We at first thought that the bars only melted because they had been previously calcined by the blast from the tuyère, which is very dense.[d] We still believe that that cause contributed at the least a great deal to the speed of the phenomenon; but after putting in through the throat similar bars, which were placed vertically and descended with the charges, we found upon removing them that those which had descended ten or twelve feet into the furnace had actually been melted at their lower extremities. Now at least ten to twelve feet more were necessary before the bars reached the hearth of the furnace, and it was not possible that they had encountered dephlogisticated air capable of bringing about their calcination even partially. Refined iron is therefore fusible without addition, but at a temperature higher than those which we customarily produce in our workshops.

[d] *In this experiment it seems likely that the iron is melting simply from the heat of oxidation in a blast of air. In the next experiment to be described, the melting would probably actually be a result of carburization in the furnace. Though both experiments are misleading, the conclusion of ultimate fusibility is, of course, correct.*

does not [138] have all the ductility of which it is susceptible — its fracture is still brilliant and lamellar — but the chemical operation is done, and to acquire the rest of its missing ductility, the iron does not need to change its nature but requires only a purely mechanical operation. Indeed, in heating it to elongate it by means of several sieges under the hammer and in folding it over on itself to make it still longer, the molecules lose the arrangement determined by crystallization, and they are, so to speak, spread out lengthwise. Finally the bar, having become much more flexible, breaks only after several bends back and forth and presents a fracture that is dark and fibrous. These fibers are ordinarily known by the name of sinews [*nerf*]. The iron then no longer has any of the brittleness that it had in its cast state; it is soft to cut, and, although it is still not a perfectly pure substance, it is at least iron in the purest state that we know.

The facts which prove that the fibers of the soft iron [139] are only the mechanical effect of the forced elongation that is produced on the metal under blows of the hammer, or in passing it through the rolling mill, or wire-drawing plate, etc., are: (1) every time that fibrous iron is heated and softened to the point that its molecules may cede to the force that tends to make them crystallize in a certain fashion, the fibers disappear without removing any of the other qualities of the iron; (2) upon elongating the iron again by the same means, the fibers return, as does the *nerf* that had been destroyed. The fibrous state of wrought iron is therefore not a part of its nature. However, when the metal is cold short and contains siderite, it is much less disposed to assume that state, perhaps because it is then fusible and can pass more easily to a crystalline state, or perhaps because it is less susceptible to extension. If it does become fibrous, the fibers are generally shorter than those acquired by soft iron in the same circumstances.

The principal object of all the finery processes, that is to say, the conversion of the cast iron into iron bars, is therefore the result of two very distinct operations. One, which is purely chemical, takes place in the crucible of the finery and has for its aim the removal of fusibility from the iron and its return to a malleable state. That is fining properly so called. When the iron has undergone that first operation, it is customary to say that it has come to nature. The other operation, which is entirely mechanical, has the double effect of bringing about a sort of cleansing by compression and of giving the iron the form of bars, the shape in which it is most commonly used. That is hammering [*martelage*]. Of these two operations, the first is our only concern; in regard to the second, it is sufficient to have properly distinguished its specific effects.

Different cast irons are not equally easy to refine. In general white cast irons come to nature more easily than those that are gray. It does not suffice for the latter to renew their contact with the coals, but [140] it is also necessary to return them perpetually to the blast of the bellows. It is this difficulty of refining gray cast iron that results in its use only for objects that must be cast and in that state then worked over with tools.

The kind of charcoal used is not unimportant to the refining. The constant custom of the forge masters who have both blast furnaces and fineries is to send oakwood charcoal to the furnace and to reserve for the finery charcoal from beech and other white woods whose combustion is easier.

Cementation of Iron

Without leaving the metallic state, wrought iron can, by virtue of a third operation, change even more considerably in nature and acquire very remarkable new properties. Small bars of the metal are stratified in chests of clay with charcoal (to which are ordinarily added other substances the kind and amount of which vary in different factories), and after having covered the vessel and sealed it with lute, it is then exposed to the action of a lively fire of predetermined intensity and duration. After cooling, it is found that by this operation, which is called cementation, the bars have experienced at least a beginning of fusion. They are much more brittle than they were before; not only have their fractures, which are no longer fibrous, again become brilliant and lamellar — which would be only the natural effect of the high temperature that they have undergone, favored, as we will see in what follows, by contact with carbonaceous material — but also they are changed in nature and composition. They have increased in weight, and their properties are changed. This is steel.

After cementation the metal is called blister steel [*acier poule*]. Its surface is ordinarily covered with blisters, and its mass is shot through with cavities of various sizes. It is clear that during [141] the operation there is released from the iron an elastic fluid which has lifted up parts of the metal and that the metal has become fluid enough to permit this release, which would merit the name of effervescence if it were more copious.

In the state in which the bars then are, they cannot yet be employed for their intended use. It is necessary to forge them, that is to say, to compact them by blows of the hammer, and to hot-weld together the parts that the bubbles of effervescence had separated. This mechanical operation changes the texture of the parts of the metal as it changes that of soft iron, but it does not give fibers to the steel, it gives it *grain*; that is to say that after having been forged, the fracture of the steel is no longer shiny and lamellar, it is gray and granular.

In being converted into steel, not only does iron become harder and more brittle, but it again becomes fusible. In the same fire it acquires more softness than soft iron, and when the temperature is raised, it is made to enter into a true fusion. It thus assumes some of the characteristics of cast iron. We will see by what follows, however, that these substances have some essential differences.

The principal property of steel, and the one for which iron is given this new character, is that on being first heated red hot up to a certain point, then quickly cooled by a sudden immersion in cold water, it develops a hardness by virtue of which it cuts glass and all substances in Nature except stones that strike sparks, which are given that name only because they have in their turn hardness enough to cut steel itself.

This immersion of red-hot steel, which is given the name of quenching [*trempe*], does not in any way change the nature of the metal: quenching does not alter the composition of the steel and transmits to it such great hardness only by an operation that is still, so to speak, a mechanical one. The rapidity of the [142] cooling, or the sudden retreat of the caloric matter [*matière de la chaleur*] that held the molecules of the red-hot steel a certain distance apart from each other, leaves a greater energy to the force that tends to bring them together. These molecules join together by virtue of a greater accelerating force, or rather one that is less restrained. They come closer together, and they acquire more adherence to each other; but, since the force which causes this coming together only acts at insensible distances, the entire mass does not participate in this condensation. It even retains greater volume and a lesser density, and it is much more fragile. That is to say, in hardened steel the contacts of the molecules are closer together but less numerous, the secondary elements are harder, and their adherence is less.[e] Finally (to use an analogy that we employ only to simplify understanding and which we do not wish to be taken rigorously), hard steel is somewhat like sandstone, which is composed of grains of quartz that are very hard and can cut hardened steel, but whose adherence to each other is much less considerable, so that they can be separated by slight shocks without themselves being broken up.

Whatever the value of this explanation, it is easy to prove that hardening does not in any way change the composition of steel, because if hardened steel is tempered [*recuit*], that is to say if it is

[e] *This theory to account for the hardness of steel on the basis of contacts between the parts is essentially that of Descartes as modified by Rohault in his* Mémoires de Physique (*Paris, 1671*), *and more or less adopted by Réaumur in 1722. That it had not changed since shows how little attention had been paid to structure in the eighteenth century.*

heated to the point of redness, which puts the molecules back at the spacing that they had immediately before quenching, and if it is then allowed to cool slowly, it no longer takes on the hardness that characterizes quenched steel, but it remains soft steel. It can be quenched and tempered as many times as may be wished, without its showing in the whole succession of operations the least alteration that must be attributed specifically to the quenching.

It is not necessary for hardened steel to have the same degree of hardness for the different uses that are made of it. The custom in this regard is to quench it very hard at first, then [143] to soften it afterward to the degree that is desired, by tempering. To do that it is heated on coals or on a mass of red-hot iron. In passing through different temperatures, steel assumes successively the following colors: pale yellow, golden yellow, purple, violet, light blue, and the color of water. Heating is stopped at whatever color is known from experience to mark the acquisition by the steel of hardness suitable to the object, and the steel is then plunged into cold water. These colors that hardened steel takes on its surface by tempering are the effect of a beginning of calcination. Because they appear on steel in a manner not only more marked but also at much lower temperatures than on iron, it follows that steel is, like cast iron, much more combustible than iron.

Cast iron and steel are analogous in still other ways. For example, when steel is heated while hot in open air, in a forge or in the focus of a burning glass, it burns while throwing far off sparks, continually succeeding one another. The sparks have exactly the same shape as those that gray cast iron throws off under the same circumstances. It is because of this facility of burning that the workmen who forge steel sprinkle it with sand, which by melting makes a varnish under which the metal is protected from contact with the air.

M. Rinman has observed that if a drop of nitric acid[f] is

[f] Acide nitreux: *literally nitrous acid, but we have adopted modern terminology to avoid confusion. For the other mineral acids — acides vitriolique et marine — we retain the contemporary English equivalents, vitriolic and muriatic acids, respectively, instead of sulphuric and hydrochloric acids.*

put on the steel, it leaves a black stain on the surface after it has corroded the metal, while on soft iron the spot where acid has been put does not change color.[g] The same thing happens with cast iron, and in that case the stain is blacker as the cast iron is more gray. This observation proves at least that the parts that constitute cast iron and steel are not entirely soluble in acids. But these two substances have essential differences. We have seen that cast iron, if it is gray, becomes white by the effect of a high temperature and approaches nearer to the state of wrought iron. On the contrary, [144] steel which is kept from contact with air, or with any materials that could exercise some action on it, withstands the highest temperatures and may be exposed to them for a long time without undergoing any change in its nature.[h]

Nevertheless, that which heat cannot produce by itself it can achieve with the aid of atmospheric air. Each time that steel is heated in free air in order to forge it, its surface loses its steely properties. If this operation is repeated often and the steel is each time refolded on itself in order to carry the surface parts to the center, in proportion as these are altered, the steel is made to lose its character little by little. It is finally brought to the state of wrought iron, its fracture becomes fibrous and the stains formed by acids are less black, so that one would have the right to conclude that the substance that gives iron the quality of steel is a combustible material which, like all such, needs contact with dephlogisticated air in order to burn, and that the product of that combustion is volatile, since it disappears.

When, during cementation, the temperature is carried too high

[g] *Sven Rinman's paper, "Rön om etsning på järn och stål," was published in the K. Vetenskaps Akad. Handlingar (Stockholm), 1774, 35, 1–14. A full English translation by Alexander Chisholm is given in C. S. Smith, A History of Metallography (Chicago, 1960), pp. 249–55. It is an important paper, the first to suggest that carbon (in the form of plumbago) was responsible for the differences between steel, cast iron, and wrought iron.*

[h] *The authors are perhaps right that heating will make iron more white but hardly that it approaches wrought iron more rapidly than steel. This misunderstanding of the nature of white iron was to lead to confusion for a long time.*

or sustained too long, the steel reaches a true fusion. The masses that result from this after cooling are still more fragile and more fusible than is steel of good quality, their fracture is black and spongy, and the stain that acids make on their surface is darker. The fusibility of these masses and the slight adherence between their parts make it impossible for them to be forged by ordinary means; they crumble away and are dispersed by the hammer. They burn in free air with more facility than steel, and the sparks that they throw off while burning are more multiplied. They possess all the distinctive characteristics of steel, but to an excessive and impracticable degree. They harden in quenching, and perhaps even more than steel, but they crack in that operation, and most often they break into pieces. As they cannot be forged, no one [145] knows how to cold-work them, that is to say, to bring closer together the parts that the effervescence of the cementation has separated too much, and their texture is too loose to be employed in the applications for which steel is intended. In short, no profit is drawn from this substance, at least in France; but we will see from what follows that this is probably a result of lack of knowledge of the nature of this substance, possibly equal to steel in value, and of not having employed processes that suit it.

The sequence of operations on iron of which we have just given a succinct description is not generally followed everywhere. In some parts of France and in Germany, the steel is derived immediately from the cast iron by a special refining operation, without making it pass beforehand through the state of wrought iron. It is then called natural steel, while that which results from the operation that we have described is called cemented steel. We have chosen to describe the sequence of operations by means of which iron passes, in a more marked way, through the four states in which we wish to consider it, and in the description that we have given, we have intended only to make known the operations that are necessary to make it pass from one state to the other. We are now going to cite the discoveries of M. Réaumur on the nature of iron considered in its different states, and then those of M. Bergman.

Extract from the Researches of M. Réaumur

In the immense body of work by M. Réaumur on iron, this diligent physicist proposed two distinct objectives in succession: the first was to discover a reliable process for making cemented steel; the second was to soften products of cast iron and, without deforming them, to give them a ductility approaching that of wrought iron.

Having at first no prescription for cementation and working from the processes for pack-hardening that are in the [146] hands of all workers of iron, he tried to cement that metal with each of the substances entering into the different compositions which serve for that operation, as well as with a host of others, all used alone and unmixed. He found that among the great number of materials which he put to the test, only wood charcoal, coal, charcoal from old burned shoes, soot, horn, and pigeon dung have the faculty of converting iron into steel when used by themselves.

These experiments enabled him to make a remark which we have also had occasion to verify and which appears important enough to us to be reported, although it may be foreign to our subject. This is, that if soft iron is cemented in crushed glass, which necessarily melts during the operation, then after cooling, the iron, which showed no alteration other than to soften more (if it was not already completely softened), comes out of the cementation perfectly clean and free from scale, because the glass dissolved all the parts of the surface that had suffered a beginning of calcination and were consequently soluble in that menstruum.

It already followed, therefore, from the experiments of M. Réaumur that among all the substances in which iron may be cemented, it is only those that are carbonaceous, or that can be converted into charcoal during the operation, that have the faculty of giving to iron the characteristics of steel. Réaumur deduced that result, but chemistry was not then sufficiently advanced for him to be able to perceive distinctly the nature of the change undergone by the iron, and he contented himself with drawing this conclusion, perhaps good enough for his time but too vague today; that the cement transmitted some sulphurs and salts to the metal, changing it into steel.

He then tried to mix different substances with the charcoal in order to identify those which could favor [147] its action in cementation, and he recognized that all were either indifferent or harmful, except pure marine salt and sal ammoniac, from which he thought he perceived good effects.

He recognized that cementation should not be continued for too long a time, that it was more economical to give the fire a greater intensity than to continue it for too long a time, and that bars half as thick did not require as much as half the time for cementation to the center as was necessary for bars twice as thick.

He sought to know which kinds of iron were by their nature the most suited for conversion by cementation into excellent steel; and although he seems to have recognized that the fibrous state of the fracture of wrought iron is only the result of a forced extension, nevertheless he proposed it as the character of iron most amenable to cementation.

He noticed the bubbles with which blister steel is permeated, and this would reasonably have led him to the discovery of the composition of iron in its different states if the theory of effervescence had been known then as it is today. He perceived the increase in weight that iron acquires during cementation and the reversion of steel to the state of wrought iron by means of successive heatings. These things he had to attribute, and did indeed attribute, to the dissipation of the sulphurs and the salts that the iron had taken up in order to become converted into steel.

He then examined the result on steel of the operation of quenching. He recognized very truthfully that it consisted only in a speedy and sudden cooling of steel dilated by heat; that substances like water, which need more heat to acquire the same temperature, were the best for this purpose, and that all the prejudices about more or less favorable waters were without basis.

Nevertheless he did not regard hardening as a [148] mechanical operation, and he believed that the sulphurs and salts detached by means of heat from the elements of the iron, at least in part, and surprised by a too quick cooling, did not have time to recombine with the elements and served them as a glue. By that means he accounted for the hardness and increase in volume of the tempered steel. Thus, after having proved that heat affects the composition of steel, he believed that heat gave rise to a sort of decomposition, an inconsistency that must be pardoned because of the state of physical knowledge at the time.

M. Réaumur did not content himself with having found good processes for cementation, but he also sought processes by which cemented steel could most quickly be restored to the state of wrought iron. He found that if steel was cemented in incombustible powders, like clay, chalk, quicklime, or slaked lime, or crushed glass, it lost its character; but of all the substances that he put to experiment, none appeared to him more proper for the purpose than the powder of calcined bones, in which steel becomes perfectly soft iron and in a period three times shorter than that which is necessary for cementation. He explains that phenomenon by supposing that these substances reabsorbed the sulphur and salts with which the steel was penetrated.

In the work which M. Réaumur next did on cast iron, he recognized, as we have already said, that the fracture of that substance varies according to the amount of charcoal used in the furnace charge, yet he regarded gray cast iron as more impure and charged with more earthy substances than white cast iron. He perceived the analogy that exists between certain properties of cast iron and those of steel, such as that of being harder than iron and of being susceptible to acquiring an extreme hardness by quenching;

and especially he made that capital observation that if a bar of soft iron is plunged into a bath of gray cast iron, [149] it is soon converted into an excellent steel. But not having the knowledge that we have today on the calcination of metals, it was impossible for him to recognize how those two substances essentially differed. He believed therefore that the cast iron was steel pushed to the highest limit, altered in addition by a residue of earthy materials, and that steel was a state of iron midway between the state of cast iron and that of soft iron.

It results from this account that, according to M. Réaumur, steel differs from soft iron only by sulphurs and salts that have been transmitted to it by the substances that enter into the composition of the cement; and that the sulphurs and salts can then be removed from it, either by dissipation in free air or by reabsorption on the part of substances that have a greater affinity for them; and finally, that cast iron is only a too-cemented steel, altered in addition, especially in gray cast iron, by earthy substances of which it has not been entirely deprived in the blast furnace.

Extract from the Researches of M. Bergman

The operations of M. Réaumur had as their principal goal the discovery of certain processes in the arts about which foreigners were secretive, and the perfecting of those processes. The object of M. Bergman's researches was purely theoretical, and these researches were uniquely directed towards an analysis of iron and the causes of its different properties. That great chemist, whom the Sciences lost too soon,[1] had discovered that metals can dissolve in acids only after having abandoned part of their phlogiston, which can be translated into modern theory by saying that metals must have experienced a beginning of calcination, and be already combined with a portion of dephlogisticated air, to be soluble in acids.

He was convinced that metals contain two distinct amounts of

[1] *Bergman died on 8 July 1784 at the age of 49.*

phlogiston; a first, less adherent one, [150] which can be removed by calcination, and another of which it is more difficult to deprive them, which they conserve in the state of calx and without which they would be acids. He gave to the first the name of reducing phlogiston and to the second the name of coagulating phlogiston. The object of his first researches on iron was to find the different quantities of reducing phlogiston that this metal contains in its different states. Because he still believed that the formation of the inflammable gas that is obtained from the dissolution of iron in certain acids was due to the dissipation of that phlogiston, he proposed first to measure the volumes of that elastic fluid which are released when different irons are dissolved in vitriolic acid and in muriatic acid, and then to judge the quantity of reducing phlogiston given up in this operation from the volume of inflammable gas produced. After a great number of experiments, he found:

1. That the same iron always gave the same volume of inflammable air, whatever might be the volume of those two acids in which the solution was done and however rapidly the operation was carried out.

2. That cast iron, steel, and wrought iron dissolved in the same acid gave consistently different volumes of inflammable air, and approximately in the relation of 40, 48, and 50. We say approximately because there are some anomalies in the results, some of which conform well to the nature of irons put to experimental trial, but others of which must be attributed to variations that occurred in the temperature and in the pressure of the atmosphere during such a long set of experiments: M. Bergman does not appear to have taken account of this.

3. That cast iron from the same ore gave more inflammable air when it was less white or when more charcoal had been used in its reduction.

4. That steel always gave the same volume of inflammable air,

regardless of whether it was hardened or soft. [151] The only difference that he observed in this respect is in the duration of the process of solution, which was easy to foresee: for, the hardness of a substance forming one more obstacle to solution, this operation must be as much slower as the hardness is greater, all other things being equal.

As to the solutions of iron that M. Bergman made in nitric acid, that chemist did not come to any conclusion. This is reasonable since there are variations in his results that were impossible to explain before the discovery of the composition of that acid, and which it is very easy to account for today.

For example, whenever the solution occurred most rapidly, either because of a higher temperature, because the iron was divided into filings, or finally, because the substance being dissolved was less hard, the products in nitrous gas have always been less, because in all those circumstances this gas must have been deprived of more dephlogisticated air and must have approached closer to the nature of atmospheric nitrogen.[j] Thus, (1) cast iron, which always gives less inflammable air than pure iron by its solution in vitriolic and muriatic acids, gives on the contrary more nitrous gas than the latter metal upon dissolving in nitric acid — and that in the ratio of 33 to 28, because (since the iron is harder and lets itself be attacked more slowly) the temperature is lower during the process of solution, and the nitrous gas retains more dephlogisticated air; (2) the quantities of nitrous gas produced by the same wrought iron put in the acid as a mass or as filings are in the ratio of 29 to 15, because the solution is carried out more rapidly when favored by the state of subdivision into filings, and the temperature being then higher, the decomposition of the nitrous gas is more complete.

M. Bergman sought by another process the quantity of reducing phlogiston that iron contains in its different states. To that

---

[j] Mosette: *Lavoisier's tentative name for nitrogen.*

end he weighed the amounts of different sorts of iron that were necessary to precipitate a definite quantity of silver [152] from its solution in vitriolic acid. In only four experiments that he cites on the subject, he found that proportionately less of a certain kind of iron was necessary as that kind gave more inflammable air. We will see by what follows the reason why that agreement cannot be rigorous, as M. Bergman himself would not have failed to perceive if he had made more experiments of that kind and especially if he had examined the nature of the inflammable gases that he obtained in the first experiments, whose differences he did not suspect.

Persuaded that caloric matter entered into the composition of iron, and entered in different amounts according to the state of the metal, M. Bergman did another series of experiments on a very great number of different samples to determine the increase in temperature produced by the solution of the same weight of different irons in nitric acid. From the fact that in general the increase in temperature was less great for cast iron than for steel, and less again for that last metal than for iron, he concluded that these substances contain caloric matter increasing in the same order.

We have reason to think that the experiments that he made in this regard were not of a kind to instruct him on the results that he sought. One can see only that as the substances which he employed were harder, their solution in nitric acid was slower and the temperature that they gave to the solution was proportionately less high.

But among the results of those researches, there is one about which he said nothing in particular and which appears to us to merit some attention. Most soft iron causes an increase in temperature of 67 or 68 degrees on a thermometer divided into 100 parts from ice to boiling water, while some magnetic black oxide of iron [*éthiops martial*] was dissolved without effervescence [153] and produced an increase of only half a degree. This proves that the great heat occasioned by the solution of iron in nitric acid, heat of which the intensity is again diminished by the release of the elastic fluid that produces the effervescence, is almost entirely due to the dephlogisticated air of which the acid is composed. That air, leaving the liquid state in order to combine with the iron and put it into a state of dissolution, abandons the caloric matter that was holding it in the state of liquidity.

M. Bergman then sought to find by means of dissolution what the foreign substances are that can be combined in irons of different nature and in different states. The only substances that he found are: (1) manganese, with which we will not concern ourselves because it is foreign to our object; (2) the siliceous matter, which is generally more abundant in cast iron than in steel and wrought iron, and which, in that last metal, never attains $\frac{1}{200}$ of the total weight; (3) plumbago, which he found in greater quantity in cast iron than in steel, and in the latter more than in wrought iron, where the weight of that substance is at the most $\frac{5}{1000}$ of that of the metal.

From all these experiments, M. Bergman draws several conclusions, of which we are going to report only the principal ones.

He regards plumbago, as does M. Scheele, as being composed of fixed air and phlogiston, that is to say, as a sulphur that can combine with iron. But then the metal must have lost a part of its phlogiston. That accepted, wrought iron contains almost no plumbago but it contains more phlogiston and more caloric matter than iron does in any other state. Steel contains more plumbago and less of the two last elements than iron. Cast iron contains still more plumbago and less caloric matter than steel.

Thus, according to M. Bergman, in order to refine cast iron [154], that is to say, to bring it to the state of soft iron, it is necessary to remove or to decompose the plumbago that it contains and to give it a greater quantity of phlogiston — operations that are both done at the same time during the refining, because the plumbago decomposes, its fixed air is dissipated, and its phlogiston is carried over into the metal.

He also explains in an analogous manner the two different processes by which steel is commonly made. These consist, as we have already said, either in extracting it from cast iron or in cementing soft iron. In the first case, a portion of the plumbago that is in the cast iron is decomposed, and its phlogiston carries over into the metal. In the second case, on the contrary, the fixed air of the cementation-charcoal, combining with a portion of the phlogiston that is in the soft iron, forms plumbago that remains united with the metal. That metal containing less phlogiston and more plumbago is steel.

Finally, after having reported a few experiments on the calcination of iron in the wet way, on siderite, which, as we have already said, gives iron the quality of being cold-short and whose nature was determined by M. Meyer, M. Bergman finished the essay of which we have just made an abstract with a few observations on magnetism.

Such was the state of knowledge on the nature of iron considered in its different metallic states at the time we undertook the work of which we are going to give an account.

Account of our Researches

Although the opinions of MM. Réaumur and Bergman on the composition of iron considered in its different metallic states differ from one another in several essential points, nevertheless both physicists agree in regarding steel as a state of iron midway between that of cast iron and wrought iron. We believe, on the contrary, that it follows from their researches and from our own experiments that cast iron and steel [155] are not composed of the same elements. In order to show the special aspects of our conclusions better, we are going first to present them, and we will then report the reasons and the facts that have so decided us.

Cast iron must be considered as a regulus, the reduction of which is not complete, and which consequently retains a part of the base of dephlogisticated air to which it was united in the ore[k] in the form of a calx. Because that reduction can be pushed to varying degrees according to the circumstances, we regard that variation as a first cause of the differences that are observed in cast irons obtained from the same ore. In addition, since charcoal has the faculty of combining in substance and without change of nature with several metals, particularly iron, it appears certain to us that the cast iron in the blast furnace absorbs a greater or less quantity of that combustible and that the extent of that absorption (determined by the circumstances of smelting) is subject to variations which are another cause of the differences in the nature of cast iron.

Perfectly refined wrought iron would be that which on the one hand was completely reduced and on the other hand contained no matter foreign to the metal, not even charcoal. There is none of that kind on the market. The best iron from Sweden always retains a part of the base of dephlogisticated air that has escaped the operations of reduction and refining, and it is always modified by the presence of charcoal, in very small amount to be sure, but which it may be impossible to remove completely.

In cemented steel the iron is perfectly reduced, and in addition it is combined with charcoal that it has absorbed from the cement and which must be present in a determined quantity so that the steel may be of a definite quality. There is therefore, in our opinion, this great difference between cast iron and steel that in cast iron the metal is always poorly reduced, while [156] in steel it is always reduced completely. There is this similarity, that in both of those two metallic substances, iron is combined with carbonaceous

---

[k] *The idea that white cast iron still contained unreduced oxygen from the ore originated with Lavoisier and was to plague the literature for years to come. It was an understandable error, however, for white iron is produced whenever there is a deficiency of charcoal in the charge for the blast furnace. The idea arose out of the understandable enthusiasm for the chemical role of the newly discovered element, oxygen, which also for a time was supposed to be a necessary component of all acids.*

matter.[1] Thus in regard to its state of reduction, cemented steel is above wrought iron in relation to cast iron, while steel has no fixed relation to cast iron in regard to the [*amount of*] carbonaceous matter, because the quantity of charcoal that it must contain to be used in the arts is greater than that which is found in most white cast irons and less than that in certain gray cast irons.

Finally, charcoal being able to combine with iron in very variable proportions, which probably depend on the temperature, we believe that steel is over cemented when it has absorbed too much charcoal during cementation. Thus, over-cemented steel differs from soft iron only by the carbonaceous matter that it contains, and from steel of excellent quality, by a greater amount of that combustible.

We are going to try to prove all these propositions, and we will terminate this Memoir by some reflections on charcoal considered in cast iron and in steel and on the state of that substance as it leaves the metal.

Since the theory of phlogiston is no longer tenable after the latest discoveries on the calcination of metals and on the decomposition and reconstitution of water, the conclusions that M. Bergman drew from his numerous experiments must at least be presented in other terms.

We know indeed that metals that have not shown a beginning of calcination cannot dissolve in acids, that is to say, metals that have not combined with a certain quantity of the base of dephlogisticated air, which would have served as an intermediary for them then to enter into the new combination. When they are [157] dissolved in muriatic or in vitriolic acid, they begin by decomposing the water that weakens the acid; they take from the acid the dephlogisticated air that enters into its composition, and then they dissolve. The base of inflammable air, left behind, taking again the elastic state [*as a gas*], escapes and produces the effervescence that always accompanies those phenomena. Although the dephlogisticated air

---

[1] La matière charbonneuse, *strictly "charcoaly matter," for the element carbon* (carbone) *had yet to be christened.*

lost a large quantity of the caloric matter to which it owed the state of elastic fluid when it entered into the composition of water, it still retains much of it in the liquid state. When the dephlogisticated air leaves the latter state to combine with the metal and to bring about the calcination, it again abandons a large quantity of caloric matter, which contributes to the production of the inflammable air and to the considerable increase in the temperature of the solution that results. Thus, the inflammable gas that is obtained by dissolving iron in vitriolic and muriatic acids does not come from the substance of that metal: it stems entirely from the decomposition of the water, and its quantity is always proportionate to the quantity of water decomposed and, consequently, to the amount of calcination that has been brought about.

In the experiments of M. Bergman, cast iron on dissolving in vitriolic acid always gives off less inflammable air than does soft iron, in the relation of about 40 to 50. It is reasonable to conclude from this that cast iron does not require as much dephlogisticated air in order to enter into solution in acid as does wrought iron and, therefore, that the iron which is in cast iron is not perfectly reduced, and it retains a portion of dephlogisticated air with which it was, so to speak, saturated in the state of calx. This conclusion is only a translation of M. Bergman, and it will become even more believable with the observation that in blast furnaces the ore is only in circumstances favorable to reduction when it has arrived at the vault of the hearth. It is in that last instant alone that it receives a great enough action from the fire to [158] melt and become susceptible to reduction; but it then becomes liquid and falls at once into the crucible, where the slag protects it from contact with charcoal and its action. The time during which the reduction can operate is therefore necessarily very short, and even if the temperature were high enough, it would be surprising if during so short an interval this reduction were brought about completely. Nevertheless, we will later report other proofs of that proposition and will have occasion to show how much it accords with the phenomena.

But in the experiments of M. Bergman, steel (even that which is obtained from soft iron by cementation) also consistently gives less inflammable air than does iron, in approximately the relation of 48 to 50. Reasoning in the same manner, it would be necessary to conclude from this that during the cementation steel shows a beginning of calcination, that it seizes a small quantity of dephlogisticated air, and that it then requires that much less dephlogisticated air to dissolve in acids. This conclusion, which seemed to arise naturally and which M. Bergman has effectively drawn in the theory of phlogiston, presents, however, a striking contradiction. Because the process of metallic reduction and that of cementation are precisely the same, it was difficult to conceive how the same circumstances, after having occasioned the reduction of iron, could then bring about the calcination of this metal to convert it into steel, unless the humidity of the cement had contributed to this effect by its decomposition. It was therefore a matter of being assured first whether humidity contributed something to the changes that soft iron undergoes in its conversion into steel during cementation and then whether charcoal, perfectly dry and free from all that can produce elastic fluids, would act differently and would give results different from moistened charcoal. [159]

For this purpose, after having calcined and removed the gas from crushed wood charcoal by holding it for several hours in a covered crucible in the center of a well-burning furnace, we filled a small crucible with it. In this we placed a bar of fibrous wrought iron, originating in the royal forges of Guérigny and weighing 2 ounces 4 gros 22 grains.[m] In order to exclude all

[m] *The weight system was as follows:*

$$\begin{aligned} 1\ lb. = 16\ oz. &= 128\ gros = 9216\ grains\ [= 489.5\ gm.] \\ 1\ oz. &= 8\ gros = 576\ grains\ [= 31.2\ gm.] \\ 1\ gros &= 72\ grains\ [= 3.82\ gm.] \\ 1\ grain\ &[= .053\ gm.] \end{aligned}$$

(International Critical Tables, *Vol. 1, 1926, p. 6.*)
The pouce *which we have translated as "inch" was 1.065 English inches, 2.71 cm. It was divided into twelve* lignes.

humidity, we did not seal the cover of the crucible with lute, but after having tied it on with an iron wire, we inverted the first crucible into another larger one. This we filled with de-gassed charcoal in the same way and tied the cover on with an iron wire. We placed this crucible upright in the center of the furnace, and the fire was urged and sustained in great activity for five hours. After cooling, we found that the exterior crucible, which was very high, had not been subjected to the same temperature everywhere, and that the force of the fire that it had received in its lower parts, nearer to the grate, had been more violent than at the top. The bar used in the experiment, which was placed vertically at the center of the interior crucible, had in the same way shown very different effects in the different parts of its height. The top had preserved its form; in the middle the parts of the metal that were at its surface had melted and had run down and reassembled toward the bottom, so that the whole piece (which had not separated) was still squared on the top, with a construction in the middle, and the bottom was swollen and round.

The weight of the piece was 2 ounces 4 gros $33\frac{1}{4}$ grains; thus by cementation it had increased by $11\frac{1}{4}$ grains, that is to say, about $\frac{1}{130}$. The cast part was broken under the hammer, and it presented the appearance of a very gray, even blackish, cast iron. The upper part, broken in the same way, looked like blister steel. The lower portion, which had been entirely [160] melted, was intractable while hot and crumbled away under the hammer. When forged in a spoon of soft iron, it prevented the iron from consolidating; and on burning in the fire, it gave more sparks than does ordinary gray cast iron. Conversely, the upper portion which had not become melted was forged excellently, without cracking on the edges. On quenching, we found that it was steel of an excellent quality, only there remained at the center a little soft iron which was easily distinguished in the fracture. Finally, the portion where there was constriction was of excellent steel, especially at the middle of its length, but at its extremities it shared some of the qualities of the neighboring parts; near the top a little soft iron remained in the center, and near the bottom it cracked a little on the edges.

This first experiment perfectly answered the purpose for which it was planned, in making plain (1) that the changes which soft iron undergoes in its conversion to steel are uniquely due to the action of the charcoal and not to that of any gassy substance which heat could disengage from it; (2) that those changes are not in the nature of calcination, that is to say that in order to be converted into steel, iron does not absorb any dephlogisticated air, since there was no substance present that could furnish it, and the iron did not approach the state of cast iron; (3) that it is the substance of the charcoal itself which, in combining with the metal, augments its weight, changes the color of its fracture, occasions the black stains made on its surface by acids, gives it fusibility, and makes it more combustible in free air. In addition, the experiment shows (1) that in cementation the operation occurs in too pronounced a degree when a certain degree of temperature is surpassed, that the charcoal then combines with the iron in too great a quantity, and that the steel which results from it is too steely; (2) that when the proper temperature is produced, it is necessary to sustain it for a definite time which depends on the thickness of the bars, so that [161] the charcoal may have time, so to speak, to dissolve and to combine through to the center of the mass. This conforms to the observations of M. de Réaumur, which we have reported on pages 297–300.

It was not sufficient to have recognized that charcoal alone could bring about cementation of the iron or the conversion of that metal into steel; it was also necessary to know if humidity would produce any changes in that operation. For that we cemented some bars of the same iron with charcoal that had been de-gassed and then moistened with pure water. And, because we were almost certain that the water would be converted into vapors and entirely dissipated before the iron had attained the temperature necessary to the cementation, we used, to cement other bars, charcoal moistened with a solution of fixed alkali which would retain the water longer. The steels that we obtained by all these operations were of the same quality as those given by dry charcoal in the same circumstances. We noticed only that when the powder used in cementation was impregnated with fixed alkali, the cementation was a little less far advanced than in the others, which also accords with the experiments of M. de Réaumur.

It followed, at least from the above experiments that the humidity with which the cement can be impregnated contributes nothing to the cementation of the iron, which was easy to predict from the state of our knowledge. For charcoal being much more combustible than iron and having the faculty of removing the base of dephlogisticated air from iron when the metal is combined with it, then certainly, when the iron and the charcoal are placed together in circumstances in which they each can decompose the water in order to take possession of the base of dephlogisticated air, it is the charcoal and not the iron that must bring about that decomposition. So, in order to discover what happens in cementation and the kind of alteration that iron undergoes to become converted to steel, we conclude that it was necessary to direct all our researches toward the action of the charcoal. [162]

It was important for us first to know the increase in weight gained by iron in order to convert itself into steel of good quality. It was, however, difficult to perform the operation in rather large chests in an ordinary laboratory furnace. We judged it better to profit from the heat of a pottery kiln and to seek by preliminary trials the place that our chest should occupy in the kiln so that, taking their volume into account, the temperature sustained during the whole of the time of the baking of the pottery would be capable of producing steel such as we wanted.

Having found that location, we loaded a square chest with four little bars of iron from the royal forges of Guérigny, employing pure charcoal for cement. After the cementation, the steel was found to be of very good quality. It forged very well, and it took a very good grain on hardening. As to the increase in weight, the following table shows what had taken place in the four little bars of the same box.

Number of the bar	Weight of the bars before cementation			Increase in weight	Ratio of the increase to the total weight
	ounces	gros	grains	grains	
1.	4	7	$38\frac{1}{8}$	$15\frac{7}{8}$	1/179
2.	5	3	$2\frac{1}{8}$	$18\frac{1}{8}$	1/170
3.	5	2	13	$17\frac{1}{4}$	1/176
4.	5	1	$50\frac{1}{8}$	$16\frac{5}{8}$	1/180

All the little bars in this experiment had been taken from a single large bar. They were then trimmed with a file on the four faces in order to remove those parts of the surface that could possibly have undergone a beginning of calcination in the forge, and which would have been capable of giving dephlogisticated air on being reduced by cementation, [163] and would thus have altered the precision of the weighings. In spite of these precautions these results must be regarded not as absolute increases in weight but as differences between the increase produced by the absorption of the charcoal and the losses occasioned by the complete reduction of the metal, that is to say, by the departure of the small portion of dephlogisticated air that is always found even in the softest iron, as we will see below.

Indeed, in another experiment, we cemented in the same chest four small bars. One of these had been taken from a bar of Swedish iron, and the three others came from Swedish sheet iron that had been forged and converted into bars. The three latter bars contained, in the interior of their masses, the parts of the iron that had been exposed to atmospheric air while hot and had undergone a beginning of calcination. These parts, on being reduced during cementation, had to lose more dephlogisticated air. Also the increase of weight which they received is less than that of the first bar, as can be seen in the following table. [164]

Kind of bar	Weight of the bars before cementation			Increase in weight	Ratio of increase to total weight
	ounces	gros	grains	grains	
Bar of Swedish iron	5	6	40	22	1/152
Bar of sheet iron 1.	5	0	$10\frac{1}{2}$	$15\frac{1}{2}$	1/193
" 2.	4	7	$14\frac{3}{8}$	$15\frac{3}{8}$	1/184
" 3.	4	7	$22\frac{1}{2}$	$15\frac{1}{2}$	1/182

It was thus verified that in cementation iron absorbs and, so to speak, dissolves some of the charcoal, which increases its weight. By that alone one could have explained why, in M. Bergman's experiments, steel that contains a little less metallic matter than an equal weight of soft iron produces less inflammable air and decomposes less water than the latter metal in order to dissolve in acids, had the difference not been consistently too great.

But we have just seen that the quantity of metal which is greater in soft iron than in steel of equal weight is about $\frac{1}{180}$. According to the experiments of M. Bergman, the difference in the products in inflammable air given by iron and steel made from the same iron is at least of $\frac{1}{24}$, and sometimes $\frac{1}{16}$, and even of $\frac{1}{8}$. It was necessary therefore to have recourse to some other circumstance that might have escaped the sagacity of that chemist in order to explain both the difference in product and the irregularities in that respect which are found in the results of his own experiments.

First, the measure of the volumes of the elastic fluids can be considerably altered by variations in the weight of the atmosphere and by the temperature of the laboratory. M. Bergman not having made mention of the precautions that he may have taken in order to avoid these two sources of error, we thought it to be necessary to repeat the solutions of iron and steel in vitriolic acid. But, because in the great number of experiments which we proposed to do on that object, we were not sufficiently in control of the conditions to keep them constant, we believed it necessary to assure ourselves by preliminary researches of the amount by which inflammable air dilates through changes in temperature in order thereby to reduce all results to a constant atmospheric pressure and a single temperature.

For that purpose, after having blown on the end of a well-calibrated glass tube a ball [165] about three inches in diameter, and after having bent the tube near to the ball, we made two marks on the tube. Then, by weighing the quantities of distilled water necessary to fill that sort of flask [*matras*], first up to one, then to the other of these marks, we were able to graduate the tube into one-thousandths of the capacity of the flask.

That done, we filled the flask with the gas whose dilatability we wished to measure. We immersed into the water of a hydro-pneumatic apparatus, in such a manner that the end of the tube was also plunged in, and we let gas escape, until with the flask at the temperature of the water, the surface of the water in the tube was at the zero of graduations, and that at the same time the zero point was at the level of the surface of the bath. By that means we had a known volume of gas which was subject to no other pressure besides that of the atmosphere and at a temperature that was known by a thermometer actually immersed into the bath. Then, always keeping the end of the tube in the water, we immersed the ball into another vessel full of hotter water, which we raised so that when the gas took the new temperature the surface of the water in the tube was again at the level of the water of the first apparatus. This gave us on the tube the quantity by which the gas was dilated (at the same pressure) by an increase in temperature that was indicated by the thermometer in the second bath. Finally, dividing the total dilation by the difference in the degrees of temperature of the two baths, we found how much the gas used in the experiment dilates per degree of increase in temperature.

In doing this experiment for atmospheric air and for inflammable air evolved by the solution of iron in sulphuric acid, we found that (under the same pressure, and per degree of a thermometer divided into 80 parts from melting ice to boiling water) [166] atmospheric air dilates 1/184.83 of its volume, and inflammable air dilates 1/181.02.[n]

On the basis of this we were in a position to correct the errors in our experiments introduced through differences in temperature: as to those that come from the differences in the weight of the atmosphere, we corrected them according to the law that elastic fluids are all compressible essentially in inverse ratio to the compressing weight.

The following table presents the results of our experiments on the solution of different kinds of iron. Each one of them was made on 100 grains [= *5.3 gm.*] of metal. The volumes of inflammable gas are expressed in ounce measures, that is to say, in volumes of one ounce of distilled water [= *31.2 cc.*]. The first column presents those volumes, as they have been directly observed, and in the fourth we have reduced them to the temperature of 12 degrees and to a pressure of 28 inches of mercury.

The volumes of inflammable air that different irons evolve through their solution in dilute vitriolic acid

Quality of iron	Ounce measures (observed immediately) ounces	Height of barometer in. lignes	Height of thermometer degrees [Réaumur]	Ounce measures (reduced) ounces[o]
1. Gray cast iron from Guérigny, in very little pieces; rapid solution	69 4/5	28 4 1/2	9 3/5	71.67
2. The same cast iron in a single piece; slow solution	70 1/5	28 3	11	71.22

[n] *These numbers are equivalent to 1/231 and 1/226 per degree centigrade respectively, about 30 per cent larger than the currently accepted values based on ideal gas behavior, 1/298. The discrepancy is probably due to the presence of water in the apparatus, for the change with temperature in the partial pressure of water in a saturated gas is 0.3 of the gas-law expansion at 16°C. The apparatus is not unequivocally described, but the tube was probably bent through an angle of 360° and passed over the edge of a hot water bath vertically downward into a separate vessel of adjustable height. The measurements are among the very earliest quantitative studies of gas expansibility and reveal the nearly identical behavior of the two different gases. Charles's Law was enunciated the following year (1787), but it was not until the work of Régnault that better data became available.*

[o] *Text reads degrees.*

Quality of iron	Ounce measures (observed immediately) ounces	Height of barometer in.	Height of barometer lignes	Height of thermometer degrees [Réaumur]	Ounce measures (reduced) ounces
3. The same volume, measured eight days later	68	28	$3\frac{3}{4}$	$7\frac{1}{5}$	70.58
4. The same cast iron in three pieces	$70\frac{1}{2}$	28	$3\frac{5}{6}$	$10\frac{4}{5}$	71.74
5. Wrought iron coming from the above cast iron	$74\frac{1}{2}$	28	$4\frac{1}{2}$	$10\frac{1}{5}$	76.26
6. Steel coming from that iron, cemented by us, in three pieces	$71\frac{2}{3}$	28	$3\frac{3}{4}$	$8\frac{1}{10}$	74.07
7. The same steel in very small pieces	$72\frac{1}{3}$	28	$3\frac{5}{6}$	$9\frac{3}{5}$	74.12
8. Swedish iron, very malleable	79	27	11	14	77.90
9. Steel coming from that iron, cemented by us	74	27	11	14	72.96
10. Very gray cast iron for cannons, from the furnace of the Platinerie, Liège district	$74\frac{1}{3}$	27	$11\frac{1}{2}$	9	75.41
11. [Wrought] iron from that cast iron, a little rusty	$75\frac{1}{5}$	28	$2\frac{1}{3}$	$14\frac{4}{5}$	74.55
12. Gray cast iron from Couvin, Liège district; rapid solution	$68\frac{2}{3}$	28	$1\frac{1}{2}$	$15\frac{4}{5}$	67.51
13. Same cast iron, slow solution	$73\frac{1}{3}$	28	$1\frac{2}{3}$	$16\frac{2}{5}$	71.90
14. Wrought iron coming from that cast iron	77	28	$1\frac{1}{2}$	$15\frac{4}{5}$	75.72
15. Steel coming from that iron, cemented by us, which had a little rose at the center[p]	$74\frac{4}{5}$	28	$2\frac{1}{3}$	$14\frac{4}{5}$	74.15
16. Wrought iron from Montcenis in Burgundy, containing graphite	$75\frac{2}{3}$	28	$2\frac{1}{3}$	$14\frac{1}{5}$	75.23
17. Black cast iron from Hüttenberg	$60\frac{1}{3}$	28	5	$13\frac{4}{5}$	60.62
18. *Idem*, from Wolfsberg	60	28	5	14	60.20
19. *Idem*, from Eisenerz	66	28	5	$13\frac{4}{5}$	66.31
20. Very white cast iron, furnished by an iron merchant who did not know from what furnace it came	59	28	5	$13\frac{1}{5}$	59.48
21. Another piece, *idem*	$60\frac{1}{2}$	28	3	$12\frac{4}{5}$	59.77

[p] *For a discussion of the rose in steel fracture — once thought to be desirable — see fn. b, Chapter Five.*

In comparing among themselves the results which this table presents, [168] it may be seen:

(1) That cast irons in general all give less inflammable air on solution in sulphuric acid than does soft iron, which indicates that they are not as much divested of the base of dephlogisticated air as is the latter metal; that is to say that their reduction is not advanced as much as is that of forged iron.

(2) That white cast irons generally release less inflammable air than gray cast irons; the former, in effect, only give from $59\frac{1}{2}$ to $66\frac{1}{3}$ measures, while the latter furnish from $67\frac{1}{2}$ to $75\frac{1}{2}$, which proves that in the gray cast iron (which is obtained by increasing the amount of charcoal in the furnace charge and by augmenting the blast of the bellows) the reduction is more advanced than it is in white cast iron, for which the temperature in the furnace had been less high and the quantity of reducing charcoal less great.

(3) That the volumes of inflammable air evolved by the different wrought irons are subject to the least variation; that, nevertheless, quite great inequalities are observed among them. The iron from Sweden, which was the most malleable, the most flexible, and the best refined of all those that we employed, is also the one that gives the most inflammable air, which consequently is the one most deprived of dephlogisticated air and thus the reduction of which comes closest to being complete.

But it can also be seen from the preceding table that steel always gives less inflammable gas than does the wrought iron from which it has been formed by cementation. The difference in those products is less than was observed by M. Bergman, which comes principally from the fact that that chemist did not take account of changes in temperature or of variations in the weight of the atmosphere in measuring the volumes of the gases. Nevertheless that difference, which for Swedish iron amounts to $\frac{1}{16}$, and which [169] can vary according to the degree of refining of the iron and according to the cementation of the steel, was still too great to be attributed entirely to the excess of metallic matter that is present in the iron, since according to our own experiments [p. 45] that excess for the Swedish iron is only $\frac{1}{152}$. So, in order to know perfectly the sort of alteration that wrought iron undergoes in order to be converted into steel, it remained to discover why the difference between the volumes of inflammable air produced by iron and steel can be ten times greater than the difference between the weights of the metallic matter that those two substances contain.

In dissolving the different kinds of iron in acids, M. Bergman had found that cast iron and steel gave a fairly abundant black residue having the nature of plumbago and that in the case of soft iron there was much less of this residue, almost none. It was very probable that that black powder which is present in steel and is not encountered in wrought iron, was nothing other than charcoal that had been absorbed by the iron during cementation and, not being soluble in acids, necessarily remained after the solution. We have also had similar residues in the solution of cast iron and of steel. That powder was very abundant toward the middle of the dissolving, when it floated and gave place to a sort of black foam; but because our solutions were made hot as they continued, the black matter diminished proportionately in quantity, finally to disappear completely before the last particles of the metal were entirely dissolved. We then assured ourselves by direct experiments that that matter is insoluble in sulphuric acid even on boiling. It had to be, therefore, that it was dissolved in the inflammable air.

M. Berthollet had already shown that charcoal was soluble in that elastic fluid.[q] We suspected therefore that inflammable air contracted and diminished in volume by that solution and that it was likely [170] because of that contraction that the inflammable air which is removed from the dissolution of iron in sulphuric acid is always nearly two times less light than that which is obtained

---

[q] *I.e., that combination would occur on passing hydrogen directly over hot charcoal. The hydrocarbons that result when annealed steel is dissolved in acids come from iron carbide, not carbon itself.*

directly through the decomposition of water. That conjecture has been verified by the experiments which M. Berthollet did alone and which he reported to the academy. He showed (1) that the inflammable air which has been within reach of dissolving charcoal and of which the specific gravity is greater, needs much more dephlogisticated air for its combustion than if it were pure; (2) that after estimating the volume of the portion of dephlogisticated air that is used in the combustion of the carbonaceous matter from the volume of fixed air which results from this combustion, the remainder of the dephlogisticated air must have been regarded as having been used in the combustion of the inflammable air. It is by estimating, from that remainder, the volume that the inflammable air would have had had it been pure, that we found this volume to be much greater than that of the inflammable air used.[r]

So, when the inflammable air dissolves some charcoal, which augments its total weight by that of all the dissolved charcoal, its specific gravity increases for two reasons — (1) because the mass increases; (2) because its volume becomes less.

We are therefore now in a position to explain why steel gives a smaller volume of inflammable air than does soft iron of equal weight by its solution in sulphuric acid. (1) Since it contains some charcoal that is not in the iron, it follows that (for equal weights) it contains less metallic matter. It must therefore decompose less water in order to calcine to the point necessary for its solution and thereby yield less inflammable air. (2) The inflammable air that it does produce is in contact with more charcoal than that which results from the solution of the iron, it dissolves more of this

---

[r] *This paragraph simply says that the inflammable gas resulting from the dissolution of steel contains hydrocarbons and therefore is more dense and requires more oxygen for its combustion than does an equal volume of pure hydrogen. A mole of hydrogen weighing 2 gm. requires half a mole of oxygen to form water. A mole of methane, CH_4, weighing 16 gm. and occupying only half the volume that the contained hydrogen would if free, needs two moles of oxygen in burning to CO_2 + $2H_2O$.*

combustible substance, and because of this, its volume diminishes at the same time that its mass augments. [171]

On the basis of this, we feel able to conclude both by synthesis and analysis that cemented steel differs from the soft iron from which it was made only by the charcoal that this latter metal absorbs during cementation. For, (1) when the cement is pure charcoal and the iron can absorb no other substances save charcoal, this metal is converted into steel of an excellent quality; (2) the analyses of iron and of steel only differ between themselves by a black powder that may be removed from the latter metal and that is not encountered in the former, at least not in such abundant quantity. This black powder, when it is entirely freed from iron, is only charcoal since it is soluble like charcoal in inflammable air, and the result of that dissolution produces fixed air by its combustion.[2]

It follows from this that it is not by the volumes of inflammable gas given off by iron and steel when they are dissolved in dilute vitriolic acid that one must judge the quantities of dephlogisticated air which each of these metals absorbs in order to dissolve; nor must one judge the quantity of carbonaceous matter that the metal contained by the weight of the black residues that remain at the bottom of the solution; (1) because the carbonaceous residue is diminished by all that which is combined with the inflammable air; (2) because the volume of inflammable gas has been decreased by the charcoal that it has dissolved, so that the iron contains more than would be immediately concluded from our experiments. To arrive at exact results in this regard, it would be necessary to analyze the inflammable gases let off by the dissolution of the irons, that is to say, first to find by combustion the quantity of charcoal that each [172] of them holds in solution and then to seek the volume that it would have occupied had it been pure.

[2] M. Rinman had observed that the inflammable gas which comes from the solution of steel gives more fixed air by its combustion than does that which results from the dissolution of soft iron.

It is perhaps not useless to observe here in passing that the preceding researches bring at least partly within the reach of reason the loss of weight that one of us observed in his experiments on the composition of water.[s] For, since the inflammable gas that he employed had been released by the solution of the iron in vitriolic acid, it must have contained charcoal and produced fixed air by its combustion. This fixed air, removed to the receiver through water, must have combined with this liquid and caused a loss of weight which it was then impossible to suspect, and which he was not able to guard against.

Before going farther, we will try to explain the principal differences that exist between the properties of iron and those of steel.

(1) Following the observations of M. Rinman, acids produce a black stain on the surface of the steel and do not produce the same effect on soft iron, because in dissolving the metallic parts of the steel they leave the charcoal, which they cannot dissolve, exposed.

(2) The properties of steel are altered in proportion as it is given successive heatings and as it is folded upon itself in forging. After a great number of similar operations, it could be reduced entirely to soft iron, because the charcoal that is at the surface is burned by its contact with atmospheric air, and by dint of renewing the surfaces, it is finally removed almost entirely.

(3) When steel is heated strongly and made white hot, it burns in a manner that is not the same as iron: it shoots afar noisy sparks that follow each other perpetually and break up in the air, because the charcoal that enters into its composition, burning rapidly, produces [173] sudden small puffs of fixed air and gives rise to small explosions. These explosions detach from the surface of the bar some

---

[s] *This is a reference to Monge's famous paper on the combustion of "inflammable gas" and "dephlogisticated air" (i.e., hydrogen and oxygen) in closed vessels. The work was done in June and July 1783 and published in the* Mémoires *of the Academie des Sciences for that year, which appeared belatedly in 1786.*

molecules of steel, which burn in the air and divide again for the same reason.

After what precedes it would be almost useless to observe that overcemented steel — which has acquired that quality only by being in contact with charcoal at a higher temperature and which has increased its weight on cementation more [*than properly cemented steel*]; on the surface of which acids leave a blacker trace; which, on dissolving in vitriolic and muriatic acids leaves a greater residue of black insoluble matter; which is more fusible; which burns more freely in free air; and which on burning sends off more numerous sparks, etc. — this overcemented steel is nothing other than iron that has absorbed by cementation an amount of charcoal greater than that which it must have to be still capable of being welded when hot and of enduring the hammer without being dispersed in fragments.

Up to here we had only found the theory of cementation, we still did not know the causes of all the varieties that are observed in cast irons. In truth we knew already that both white and gray cast irons can differ among themselves by the degree to which the reduction of the metal is carried since they disengage different quantities of inflammable gas when they are dissolved in acids, but it remained for us to discover the causes of numerous properties by which the gray cast irons differ from the white ones and, in particular, the cause of the differences of color that those two kinds of substance present in their fractures.

Now we have already remarked at the beginning of this Memoir the analogy that is present between the properties of gray cast iron and those of steel. Gray cast iron is stained black by acids; like steel, it leaves a black residue after solution in acids, [174] sometimes even a more abundant one; it burns in free air, throwing off sparks; and finally it is capable of being hardened by quenching. That analogy alone would suffice in order to conclude that in that substance, as in steel, the iron is combined with a certain quantity of charcoal, sometimes even in a greater amount than in the latter metal, but what proves it incontestably is the faculty that gray cast iron has of cementing soft iron that is immersed in it when molten and converting it into steel. Independently of the observation of the phenomenon that is made daily in the furnaces where gray cast iron is cast, we have ourselves had several occasions to verify it.[t] Whenever we heated some cast iron at the forge, the pokers that we were using to bring it back into the fire became cemented in the parts that touched the cast iron. They became susceptible to hardening by quenching and displayed the grain of steel in their fracture. We also did a direct experiment on this subject.

In an individual crucible, we put a bar of soft iron from Guérigny with an equal weight of gray cast iron from the same forge. We had covered it all with crushed glass, which, on melting, should keep the two metals protected from contact with atmospheric air. This crucible was exposed to the action of the fire of a good furnace for five hours. After cooling, we found that the cast iron had been melted, that it had become whiter than before, and that it had acquired ductility. As for the bar of wrought iron, it had kept its shape, it had touched the cast iron only in a few parts on three of its faces, and it was not even joined everywhere with it; but wherever it had been in contact with the cast iron, it had become steel of good quality, while the parts remote from contact were still only soft iron.

It follows, therefore, that in gray cast iron (which may be [175] considered as a quite good cement and which has the faculty of transmitting charcoal to soft iron in order to convert it into steel) the metal is united with a fairly large amount of charcoal that it

---

[t] *Although the authors regard it as a curiosity, the intentional making of steel by the immersion of wrought iron in molten cast iron was an old process, perhaps an extremely ancient one, though the first printed description of it in a European language is in Biringuccio's* Pirotechnia *(1540). See this volume, Chapter Two. Cast iron probably had a transitory existence in the intermediate stages of producing "natural" steel in a hearth somewhat before it was intentionally made as a separate product.*

took up in the blast furnace. When the cast iron is liquid and sufficiently hot, the charcoal in it is in a veritable state of solution, since it is distributed in an obviously uniform manner in the whole mass in spite of the differences in the specific gravity of the two substances, and especially since it acts upon soft iron that is presented to it in the same way that salt dissolved in water distributes itself in new water that is added to the solution. The cast irons that are called "mottled" (the fracture of which is not of a uniform color, but is composed of both white and more or less gray cast iron) are in this state because their reduction in the furnace did not occur in the same manner all over, and because they had not been held fluid enough or long enough in fusion for the solution of the charcoal to have had a chance to become uniform.[u]

There are therefore two principal causes of the variety in the cast irons: the first is the quantity of dephlogisticated gas that remains united to the metal, depending on the degree to which the reduction had been carried in the furnace. The less dephlogisticated air that remains, the more the cast iron approaches the nature of soft iron: it is the base of the dephlogisticated gas that renders the white cast iron fusible, gives it fragility, and communicates to it the hardness by virtue of which it is intractable to a tool. The second cause of the variety is the quantity of charcoal that the cast iron was able to absorb in the blast furnace. It is the charcoal combined with the iron in gray cast iron and in black cast iron that gives them their colors, and it is charcoal that renders them generally more fusible than white cast irons at equal degrees of reduction, that forms the black residue that they leave behind when they are dissolved in acids, and, finally, that gives them the principal characteristics of steel. [176]

It could be objected that, since the charcoal contained in cast iron

---

[u] *The mottled texture, of course, derives from the fact that the silicon content of such irons is too low to cause profuse nucleation of the graphite, which consequently grows in localized patches in a general matrix of white iron.*

has the faculty of contracting inflammable gas into which it is dissolved, it would be possible to explain by that alone the observed difference between the volumes of inflammable gas disengaged by cast iron and by [*wrought*] iron on being dissolved, without having recourse to a defect in reduction. The cast iron would then not necessarily contain dephlogisticated air and would be nothing other than steel, the overcementation of which had been pushed more or less far. We would agree, and we have already said that this observation is a reason for believing that the quantity of dephlogisticated gas that is still present in the gray cast iron is not quite as great as might be concluded from the difference in the production of inflammable gas; but if one notices that the whitest cast irons (which do not obviously contain carbonaceous matter and which leave no black residue on solution) are precisely those that disengage the least inflammable gas, then one will be forced to admit that the white cast irons at least already contain dephlogisticated air, by virtue of which it is not necessary that they decompose as much water or that they disengage as much inflammable gas in order to be rendered soluble in acids.

As to gray cast iron, it is known that when it is held in fusion for a long time at a very high temperature, sheltered from contact with atmospheric air and from all materials that could furnish dephlogisticated air and cause combustion of the charcoal that renders it gray, it loses some of the properties that it possessed before: it then becomes less steely, its fracture becomes whiter, it acquires ductility, and it approaches more the nature of wrought iron, which is infusible at similar temperatures, while cemented steel, which similarly contains charcoal, can sustain the highest temperatures when sheltered from contact with atmospheric air without undergoing any apparent alteration. [177] In that operation gray cast iron therefore loses the charcoal that it was before in condition to transmit to the soft iron. Now how could the charcoal (which is unalterable under the greatest fire) disappear, especially as it is already combined with the iron, if it did not encounter in the gray cast iron a remainder of dephlogisticated air capable of bringing about its combustion?[v]

Thus, cast iron is a metal of which the reduction, though more or less advanced, is not carried far enough for the iron to have ductility. It may, moreover, contain a larger or smaller quantity of charcoal that it may have absorbed in the blast furnace or be almost entirely deprived of that substance. As a consequence, cemented steel (which conversely is always in a state of complete reduction and, in addition, necessarily contains charcoal) is not a state of iron midway between cast iron and refined iron as was believed until now.

Finally, perfectly soft iron, if any of this kind existed, would be a pure metal, entirely deprived of both the base of dephlogisticated gas (with which it was combined in the ore or in the cast iron) and of the charcoal that it would have absorbed in the furnace during the reduction. But the principal operations of refining, which all have as their aim to purge it of these two foreign substances, are not capable of such great precision that the purification may come about completely: the best irons from Sweden always contain quantities, in truth very small, of dephlogisticated air and charcoal. In fact, the best refined iron from Sweden still contains charcoal, for it always leaves a light black residue at the bottom after it has been dissolved in acids, and it contains dephlogisticated air, since if it is cemented with charcoal in order to convert it into steel, the blister steel that is the immediate product of the cementation is always bloated and shot through with hollow blowholes of various sizes, to the point that before employing it for any purpose whatever, it is necessary to begin by forging it, to work it under the hammer in order [178] to bring together and weld together the parts of the metal that the effervescence had separated. This effervescence is evidently the effect of the disengagement of the

---

[v] *This is a good theory, but the observation it is designed to explain is erroneous: cast iron does not lose carbon on mere heating, though the characteristics of its fracture may change.*

fixed air formed by the combination of the charcoal of the cement with the little dephlogisticated air that the refined iron still retains.

Explanation of the Processes that are Followed in the Forges to Make Iron Pass Through its Different Metallic States

On the smelting of ore

In order to charge a blast furnace, definite volumes of ore, flux, and charcoal are thrown in at the same time through the upper opening. Then, as a result of the combustion that takes place in the hearth, the surface of the charge sinks in the furnace, and the operation is repeated when there is room for a new charge. The amounts of the three substances that ordinarily compose a charge vary (1) according to the nature of the ore, (2) according to the state of the furnace, and (3) according to the qualities that it is proposed to give to the cast iron. The principal object of the flux is to contribute to the fusion of the gangue and to facilitate the access of the charcoal to the metallic calx that it encloses. When the gangue is siliceous or argillaceous, the flux is ordinarily of calcareous earth and then it is given the name *castine*, from the German word *Kalkstein*, which means "limestone." When the gangue is calcareous, clay is employed as a flux and it is called *arbue*. Finally, there are several ores for which no flux is used because their gangues, being composed of different earths, are fusible with no addition.

The use of charcoal in the furnaces has two distinct objects: the first is to cause by its combustion a temperature high enough to ensure the fusion of the [179] gangue, and the second is to begin the reduction of the metallic calx by extracting from it a greater or less part of the dephlogisticated air that enters into its composition. As the charcoal divides in different ways in order to achieve these two objects, the cast iron changes its nature. For example, when a furnace is producing white cast iron with quite advanced reduction which is consequently easy to refine, if the blast at the tuyère is increased (either by giving a greater opening to the blast pipe or by accelerating the working of the bellows) without changing the amount of charcoal in the charge, then the temperature of the hearth is raised because a greater quantity of the charcoal is caused to burn and less free charcoal remains to serve for the reduction of the calx. More material must therefore be tapped, since the charges descend more quickly, but the cast iron will be less reduced, and its refining, which consists of the complement of the reduction, will be rendered more difficult. It is therefore necessary, when changing the blast in the tuyère, to change the amount of charcoal in the charge.

If at the same time that more blast enters at the tuyère, the amount of charcoal in the charge is increased in greater proportion, not only is the metallic reduction pushed further because of the greater quantity of free charcoal that has been introduced, but also, since the affinity of the iron for the charcoal is augmented by the increase in temperature, the metal combines with that fuel and bears some of it away into the bath in the crucible, and the cast iron that results is gray.

Although this cast iron in general is better reduced than the white cast irons, its refining is much more difficult than that of the latter because the operation does not consist only in achieving the reduction, but it is also necessary to burn and dissipate all the charcoal combined with the cast iron. This requires that the cast iron in the finery be frequently returned to the bellows' blast and that the surfaces of the metal in contact with the air be perpetually [180] renewed. Thus the masters of the forges always cast iron that is destined for fining as white cast iron, and they only cast as gray iron pieces such as naval cannon and [*water*] conduits that must have a little flexibility and must be subsequently worked over with a tool.

It can therefore be seen why either white or gray cast iron may be made at will from the same ore by varying only the quantity of charcoal in the charges and the blast of the bellows. However, these two circumstances are not the only ones that must be considered when one wants to obtain gray castings, for the combination of iron with charcoal requires a certain time. The very fusible ores, which run very quickly in the crucible of the finery, are in contact with the coals for very little time and are in general less suited to produce gray cast iron than the more refractory ores.

It will perhaps be difficult to imagine that, since the reduction of the metallic calces is achieved by means of contact with coals that extract from them their base of dephlogisticated air, gray cast iron can contain some charcoal and yet not be a completely reduced metal. One might think that the reduction could only be incomplete when the quantity of charcoal is not sufficient and that free charcoal could remain to combine with the iron only when there is no longer any reduction to bring about. But it must be observed that, especially in situations where the cast iron is gray, the temperature of the hearth of the furnace at the level of the tuyère is much higher than that of the bath of metal at the bottom of the crucible. This temperature would be capable of bringing about the complete reduction of the metal, and even of giving place to an exaggerated cementation, if the cast iron were exposed there for a sufficient time; but the drops that result from the fusion of the ore only undergo it during the instant when they pass in front of the tuyère, and that instant is quite short. Only the surface of those drops may therefore be reduced and absorb charcoal. In the interior [181] the reduction is much less advanced, and there is no charcoal combined at all. Then when these drops fall into the bath that is beneath the slag and at a lower temperature, they first abandon the absorbed charcoal, the greatest part of which remains disseminated in the metal. Then by a kind of communication, the reduction distributes itself in the whole mass in an apparently uniform manner without making further progress, because the charcoal that is given up is not pure charcoal. It is, as we will see later, graphite, the combustion of which is more difficult, that can only bring about metallic reduction by a higher temperature than that of the bath. Nevertheless, the existence of charcoal in the gray cast iron is demonstrated by the facts, and we have as our only object in this explanation the forming of a concept of how it can be there.

M. Bergman has done some experiments on cast iron, the results of which have appeared extraordinary, though they follow naturally from our theory. For example, that chemist, after having cemented 200 pounds of gray cast iron from Hällesfors with black hematite, and another time with iron calx precipitated from vitriol and heated red hot in a crucible, obtained two ductile reguli with an increase in weight. In the first case the regulus weighed $201\frac{1}{2}$ pounds and in the second, 206 pounds. (See Experiments 90, 91.) It is evident that this cast iron contained enough charcoal at first to achieve the reduction of the metal and then to reduce a part of the ferruginous calx that served as cement for it. The product of the latter reduction is the cause of the excess of the weight of the regulus over that of the cast iron employed.

In another experiment, 200 pounds of cast iron from Leufstad melted in a crucible without addition, and urged by a very great fire, only gave 196 pounds of regulus, and this regulus was steel of excellent quality, which was stained black by acids. (See Experiment 97.) It is evident that a part of the charcoal that was in the cast iron had been employed [182] to complete the reduction and that the other, not being able to be absorbed by any neighboring substance, remained combined with the metal and gave it the characteristics of steel. As to the diminution of weight observed in this experiment, it comes from the release of the fixed air formed by the combination of a part of the charcoal with the base of dephlogisticated air that the metal still retained in the state of cast iron.

After this it is easy to explain why, when remelting pieces of scrap gray cast iron in reverberatory furnaces, if the surface of the iron has either been calcined by the fire or rusted in the open air, the metal in the interior of the pieces melts and flows into the crucible and gives a whiter cast iron, while the exterior of the pieces is refined, assumes a pasty state, and, retaining its form, stays on the bridge of the furnace.[w] For in the interior a part of the charcoal is employed to advance the reduction of the metal, which whitens the iron, but reduction is not entirely achieved because the metal flows immediately into the crucible and does not remain exposed to the action of the heat quite long enough. At the surface of the pieces, on the contrary, the metal receives a greater blast of the fire and undergoes at first a more advanced reduction, then, by contact with the metallic calx, it loses the remainder of the charcoal that was rendering it fusible, and the refining is achieved. The refined iron which thus remains on the bridge is ordinarily called a skull [carcas], and it can be converted into bars under a hammer.

On the fining of cast iron

When the cast iron is white and it contains almost no carbonaceous matter, refining consists simply in removing the last molecules of dephlogisticated air that give fusibility to the iron and that removes its ductility. This object is achieved by having the pig iron remelted at the forge of the finery where it falls [183] drop by drop into the crucible and by then stirring the bath of cast iron in order frequently to renew contact with the incandescent coals. The charcoal, in combining with the base of dephlogisticated air of the cast iron, causes further progress in the reduction that had only been started in the blast furnace. The iron ceases to be fusible at the temperature that it experiences in the finery; it becomes pasty and it can be drawn out into bars under the hammer.

If the poker of wrought iron that the finer uses for stirring the cast iron and bringing it into contact with the charcoal remains for some time plunged in the crucible, upon leaving the bath the part of the tool that has been immersed is enveloped in a more or less

---

[w] *It was observation of a similar decarburized, unmelted skin on an iron pig in a melting furnace that triggered Bessemer's interest in the effects of atmospheric air in iron refining (see his* Autobiography *[London, 1905], p. 142). Réaumur had noticed the same phenomenon in 1722 and had correctly explained it.*

thick sheath of cast iron that has come to nature. The sheath does not adhere to the poker; it can be detached from it by shock and is capable of being extended under the hammer. The wrought iron, which at the same temperature has more affinity for the base of dephlogisticated air than has cast iron, has therefore assumed here part of the role of the charcoal: the cast iron, in dividing with it the remainder of the retained dephlogisticated air, undergoes a large enough reduction to lose its fusibility in the fire of the finery and to acquire a certain degree of ductility. It is evident therefore that it is not necessary that the iron be completely deprived of dephlogisticated air in order to be malleable and that the wrought irons of commerce can differ from each other by the state to which the metallic reduction has been carried. However, the experiments on solution prove that iron is, all things being equal, proportionately more ductile as the reduction approaches completion and that the superiority of the irons from Sweden occurs because the reduction of the metal has been pushed farther in the fining.

If one considers (1) that wrought irons always contain actually very little, though varying, amounts of dephlogisticated air; (2) that white cast irons contain much more of it, and differ [184] considerably one from another in that regard; (3) that the ferruginous *éthiops martial* [iron oxide] obtained by MM. Lavoisier and Meusnier by calcining iron by means of water vapor is evidently a state of iron midway between white cast iron and calx; (4) that the iron calces themselves can contain more or less dephlogisticated air depending on the circumstances accompanying the calcination, then one is forced to conclude that iron is capable of combining with the base of dephlogisticated air in an infinite number of different proportions, not because this metal is susceptible of several points of saturation for this substance, but because in each particular circumstance its affinity for the base of dephlogisticated air is placed in equilibrium with the forces that oppose the combination. Now the first molecules of dephlogisticated air that unite with the iron to begin the calcination adhere more to the metal than those that enter in the combination later. Therefore, when it is a matter of bringing about the reduction of the iron, the more that reduction is advanced, the more it is necessary to increase the circumstances that favor it. It is therefore necessary then to raise the temperature and to use as a reducing agent a substance that has more affinity for the dephlogisticated air, that is to say, whose combination is easier. This reasoning explains and justifies the steady custom in forges of routing the charcoal from oakwood to the blast furnaces and reserving for the fineries the charcoal from beech and other white woods that are more combustible. For everyone knows that beech charcoal continues to burn and is reduced to ashes in circumstances under which that of oakwood would be extinguished; thus when it is employed in fineries it must cause the reduction to proceed more rapidly and carry it farther than does the charcoal from hard woods.

On the contrary, the purified coal that is called coke [coak] in England is much less easily combustible than the charcoal of oakwood; [185] it is extinguished in circumstances under which the latter still burns very well; it requires a denser air, or a more rapid blast. Also this [coke], which is employed successfully to begin the reduction of the ore in blast furnaces and to produce cast iron, is not proper for achieving reduction in fineries to produce wrought iron. To bring about its rapid combustion, it is necessary to raise the temperature very high and to enhance the blast from the bellows. Because these are precisely the circumstances that occasion the calcination of the iron when refining is being done with coke, then, far from advancing reduction, almost as much metal is calcined as charcoal is burnt. The English, who do not employ wood charcoal in their forges, find it therefore necessary to abandon the processes of fining that have always been followed in the countries where wood charcoal is cheap.[x]

[x] *This is an oblique reference to the puddling furnace, which had been introduced by Henry Cort only two years earlier.*

The temperature necessary to make white cast iron come to nature through contact with wood charcoal (that is to say, to reduce it to the point of being ductile and extensible under the hammer) is below that which is capable of making it begin to melt. Indeed, in certain forges the process that we have described is not followed for refining iron. The cast iron is first cast into thin plates, then they are stratified with wood charcoal, and a furnace built up and covered over with earth or with slag, much like those in which charcoal is burned. Then the furnace is lighted through a chimney which has been contrived at the center, and the fire communicates gradually to all the charcoal of the mass, which, since it is deprived of contact with atmospheric air, can only burn by removing from the cast iron a part of the dephlogisticated air contained in it and so advancing its reduction. At the end of a certain time, of a duration depending on the thickness of the plates, the metal is refined; the plates have preserved their shape and have not melted. This is what is called refining [*mazer*].[y] If the operation is stopped before it is completed and the plates are broken after having cooled, it is easy to distinguish in the fracture the parts of the iron that are near the surface of the plate and have come to nature from those at the center of the thickness that still have the appearance of the cast iron. The first parts are reduced enough to be drawn out into bars, while the reduction of the others has not been promoted.

One feels that the operation of refining as we have just described it cannot furnish directly a very refined and ductile iron. It would be necessary to make the furnace provide a greater action of the

---

[y] *The process later known as refining was quite different from this. It consisted of subjecting molten cast iron to an air blast, stopping at the point where silicon but not carbon was oxidized. The product was not used directly, but was fed to the puddling furnace in which refining was easier and more productive because of the lack of silicon. A number of processes depending on oxidation of cast iron without melting it were used for making steel, the best known being that of Uchatius, patented in 1855. (For a good discussion of both, see J. Percy, Metallurgy. Iron and Steel [London, 1864].) See also fn. h, p.144*

fire and sustain it for some time in order to give rise to a more complete reduction without, however, attaining the temperature proper to cementation, for the iron would then be converted into steel. This aim is just about achieved in the forges where refining is done by carrying the refined plates to a chafery fire in order to convert them into blooms, for there they undergo a temperature much higher than that of the fining furnace, and the contact with the charcoal carries their reduction to a sufficient degree for use.

Until now we have spoken only of the refining of white cast iron. For gray cast iron, the operation consists not only in ridding the iron of the dephlogisticated air that has resisted the action of the coals during its fusion but also necessarily in removing from it even the charcoal with which it combined in the blast furnace. This second part of the refining is in general much more difficult than the first. When the cast iron is slightly gray, that is to say, when it contains a little carbonaceous matter, two methods are ordinarily employed to remove this foreign substance from it. The first is to produce a higher temperature by increasing the blast from the bellows of the finery and to bring about the combustion of the charcoal contained in the cast iron by the dephlogisticated air that it retained. The second is to return the cast iron repeatedly to the blast from the bellows and so to occasion the combustion of the carbonaceous matter. In [187] certain forges the first object is achieved by introducing into the bath of the finery the jet of air from the bellows. That jet, in agitating the cast iron and in perpetually renewing its contact with atmospheric air, gives rise to the combustion of the charcoal that rendered it gray.

Those two processes are very feeble, and though both are customarily employed at the same time, fining is always very slow and very costly. Indeed it is at first very difficult to raise the temperature of the bath of the finery enough so that the dephlogisticated air and the charcoal that are contained in the cast iron can combine and escape from the metal. In the second place, the state in which the charcoal exists in the cast iron (of which we will speak later) makes it very weakly combustible, and in exposing the cast iron to the blast of the bellows, much of the metal is calcined at the same time as some of the carbonaceous matter is dissipated, which entails a loss.

But when the cast iron is very gray, the difficulties of which we have just spoken become even greater. There are no certain means of fining it profitably, because the losses occasioned by the calcination and by the dispersion of the metal and the expense of the fuel and labor carry the cost of fining too high for an operation that has as its aim only the returning of gray cast iron to the state of white cast iron.

However, if one had a great quantity of very gray cast iron for which there was no use other than to be converted into wrought iron, the procedure that results from Experiments 90 and 91 of M. Bergman could be followed. (We propose this nevertheless with reserve, because we have not yet had occasion to execute it on a large scale and to compare the cost that would be involved with the products that would result.) This would be to cement the cast irons, after having cast them in thin plates, with iron calx or with washed and powdered ore and to sustain a very strong fire in the cementation furnace [188] for quite a long time. By that means, the reduction of the metal would be achieved at the expense of a part of the charcoal that it contains, and the remainder of the charcoal would be employed to reduce a part of the cement, so that the result of the operation would be perfectly refined iron, of which the weight would exceed that of the cast iron employed. In truth, the expenses of that cementation would be considerable, especially if the fuel was not of a low price, but they could be offset (1) because the weight of the wrought iron would equal at least that of the cast iron, while in the best equipped fineries it is necessary to have 1350 pounds of cast iron in order to produce 1000 pounds of wrought iron; (2) because the finery could be dispensed with, and only a chafery fire would be needed to shingle the bars.

Moreover, the most advantageous use that could be gotten from a large quantity of very gray cast iron would not come from converting it into wrought iron but from making steel of it, provided however that it did not contain either siderite or foreign metals. To do this according to Experiment 90 of M. Bergman, it would be necessary to expose it to the cementation fire in closed chests, without cement. One part of the charcoal contained in the cast iron would serve to complete the reduction of the metal; the other, remaining disseminated in the mass, would convert it into steel, the quality of which would then depend on the amount of charcoal that had not been burned.

On the cementation of soft iron

There remain few things to say on cementation, on which we have already entered into considerable detail. We have seen that charcoal was the only substance that, when combined in a definite amount with soft iron, has the faculty of communicating to that metal the property of becoming hard on quenching; that, in order for this combination to be effected to a suitable degree, it was necessary for the iron to experience a certain temperature during [189] cementation, which put it in condition to absorb a sufficient quantity of charcoal, and that this temperature be sustained for a sufficiently long time so that the cementation could be effective right to the center of the bars. If therefore a cemented steel is desired that is capable of being forged easily and of being welded, it is very important to attain the temperature at which the iron will be able to absorb the necessary quantity of charcoal and not to exceed this, because the steel would then be too much cemented, and it could not be forged.

It is not that steel that is overcemented and rendered fusible by too great an amount of charcoal may not be perhaps more suitable for many applications than steel capable of being forged, for, on melting, the charcoal distributes itself in a more uniform manner throughout the whole mass, and one is not likely to find in its interior some parts that are too close to the nature of iron and that may be incapable of becoming hard on quenching, or do so less well.

The use of this material [*cast steel*] has been restricted up to the present, principally (1) because it could be employed only for objects that are cast and poured into molds and (2) because, being fusible, and not being in a hammer-worked state, it is impossible to bring together by ordinary means the parts that were separated by the effervescence accompanying cementation and to make the bubbles and the cavities that are often present in its interior disappear. But what cannot be done by shock could be reasonably achieved by pressures such as those of the coining press in mints; at least it is very probable that it is by some analogous operation that tools are made of steel, such as the cylinders for rolling mills, the hardness of which after quenching is very great and whose grain is perfectly uniform throughout the whole mass.

Moreover, if one wishes to preserve the properties of overcemented steel, which is more combustible in free air than is iron or even ordinary steel, [190] it is necessary to protect it completely from contact with the atmospheric air whenever it is heated or melted, because otherwise its excess of charcoal would consume itself in all contacts with the air, and its substance would become less uniform.

In certain forges, for example in those of Carinthia, the same gray cast iron is converted at will into soft iron or into steel. The processes that are followed for each of these objects are perfectly in accord with the theory that we have expounded. In both cases, cast iron is run into a large crucible, then by throwing cold water on the molten metal, it is hardened at the surface, and a first thin plate is lifted off. The surface of the bath is cooled anew to lift off a second plate, and in continuing the operation, the greatest part of the melt is converted into plates. That done, in order to have steel, these plates are melted at the finery where they undergo a violent heating sustained for a long time, and they are protected from contact with the air by a sufficiently thick coating of slag. In order to have soft iron, the plates are at first submitted to a long roasting, and the air is carefully renewed by means of two bellows. The result of the operation is then carried to the finery where it is treated as in ordinary forges.

Thus, in order to convert gray cast iron into steel, it suffices to heat it strongly while it is sheltered from contact with the air. By that means a portion of the charcoal that was in the cast iron is employed to complete the reduction, and the other portion that remains in combination gives to the metal the qualities of steel. In the second operation, on the contrary, the roasting in free air consumes the charcoal that is at the surface and, by communication, a great part of that which is in the interior. When the product of the roasting is then carried to the finery, the little charcoal that remains suffices for the reduction, and the metal, robbed of dephlogisticated air and of charcoal, is [191] soft iron. It seems to be advantageous in these forges to convert all the cast iron into plates, for if it is wished to obtain iron, these plates roast more easily because of their small thickness, and if it is wished to make steel, they are sooner melted, and they are buried beneath the slag before the blast from the bellows has consumed much of the charcoal that they contain.

On Charcoal, Considered in Its State of Combination with Iron, and on Leaving That Combination

We have seen that charcoal has the faculty of combining with iron and that the result of that combination must be regarded as a true solution, because these two substances are distributed uniformly in the interior of the mass in spite of the difference of their specific gravities (as is the property of solutions) and because cast iron and steel when melted transmit charcoal to soft iron that is plunged into them. This affinity of charcoal with iron evidently varies with temperature, for (1) at ordinary temperatures, these two materials have no effect upon each other, and it is necessary that they both be heated to a certain point for the solution to take place; (2) in proportion as the temperature is raised, the solution becomes more concentrated, a fact that is proven by the excess of charcoal that is taken up by [*wrought*] iron when the temperature is pushed too far during cementation, and by cast iron in the blast furnace when too much charcoal in the charge excites too high a temperature in the hearth. Thus iron is capable of being saturated with charcoal, and the quantity of the latter substance that is necessary for saturation varies with temperature.

It follows from this that if cast iron and molten steel are saturated with carbonaceous matter at a temperature much [192] higher than that which is necessary for fusion and if they are allowed to cool, the metal, whose affinity for the charcoal diminishes as the temperature is lowered, will become supersaturated and abandon some charcoal, and this sort of solution will become turbid [*se troubler*], but the condition of the mixture will be different depending upon the regimen of cooling.

If the cooling is conducted in a very slow manner, the metal will be purified, because the rejected charcoal has time to rise to the surface. As we will see in a moment, it is to this purging that one must attribute the plumbago that is found at the surface of gray cast iron cast in a large mass and that which ordinarily covers the ladles with which this material is cast into a mold. But if the cooling is too sudden, as ordinarily happens, the rejected charcoal is trapped in the metal before it has been able to disengage itself, and it finds itself disseminated in the interior and not combined.

Now the affinities of two substances always being reciprocal and iron having the faculty of dissolving charcoal, charcoal must be regarded in its turn as capable of retaining iron. In addition, every time that a precipitation is made without an intermediate, the rejected substance is always saturated with the solvent. It is thus that air rejected by water by virtue of an increase of temperature or a decrease of pressure is always saturated with water. Therefore the charcoal that had been held in solution in some pig iron and that was rejected because of cooling must be saturated with iron: it is no longer pure charcoal, but it is plumbago; that is to say, it is the same substance as that of which English pencils are made.[z]

Actually, the substance that rises to the surface of gray cast iron during cooling has all the exterior characteristics of true plumbago; it has its color, it is soft to the touch like plumbago, it leaves traces on paper, and it acts in the fire exactly like [193] plumbago. To be sure, it is most often encountered in small, very thin lamellae like mica and not in adherent masses capable of being cut into pencils. This can come from the circumstances of its precipitation, principally from the speed of cooling. However, it is also sometimes found in solid masses. We have had occasion to observe the demolition of a furnace in Champagne from which a gray cast iron of good quality had been cast (the iron having been converted to sheet iron), and we have found some debris among the stones of the structure to which were adhering massive pieces of plumbago of the thickness of 6 or 7 *lignes*, crystallized in a regular manner. Unfortunately, it was not possible for us to judge of the form of the crystals, because these pieces had not been spared during the demolition, and the crystals had been broken.

In addition, all the analyses that M. Bergman has made of the black residue that is found at the bottom of solutions of gray cast iron and of steel in acids prove that this residue is absolutely the same material as plumbago, and all those that MM. Scheele, Hjelm, and Pelletier have done on plumbago, prove that that substance is nothing other than charcoal combined with a certain

---

[z] *Although this argument is in principle correct, graphite in equilibrium with iron actually contains very little metal. At the time, however, graphite was regarded as a kind of subcarbide of iron, a belief based on studies by Scheele and Bergman who had noticed that the natural mineral contained iron. They were further confused by the fact that the residue left on dissolving steels and white irons in acid did indeed contain iron (in the form of iron carbide), unless the solution was protracted for an unusually long time to allow its decomposition. Natural graphite commonly contains about 1 per cent iron, but there must have been more in the samples that were tested to produce the effects here described.*

quantity of iron. We will content ourselves with reporting here the principal ones.

(1) Plumbago is inalterable under the hottest fire in closed vessels, and when it is calcined under a muffle, it loses $\frac{9}{10}$ of its weight, and the residue is a ferruginous calx.

(2) When it is detonated with saltpeter, it produces fixed air, and it gives a ferruginous residue.

(3) When it is distilled with sal ammoniac, this salt is sublimed into ferruginous flowers [*fleurs martiales*], that is to say, into flowers of sal ammoniac charged with iron. [194]

(4) We digested some plumbago in very pure muriatic acid; during the digestion, a little inflammable air was evolved, at first $\frac{3}{100}$ of the matter employed was dissolved, and the part dissolved was iron that we precipitated as Prussian blue with Prussian lime water prepared in the manner of M. de Fourcroy. Then, after having calcined the residue, we exposed it anew to the action of the muriatic acid, and we again obtained some iron. Finally, continuing thus to favor the solution of the iron by the combustion of the charcoal and to facilitate that combustion by the solution of the metal, we came to extract quite a great quantity of iron, which was, however, impossible for us to measure with any exactitude. The inflammable air that is obtained in this series of operations is produced by the solution of the iron in the acid and proves that the iron which constitutes plumbago is in the metallic state.

It follows first from these experiments that plumbago contains iron. What follows then proves that it contains charcoal.[aa]

(1) Plumbago revives litharge and arsenic acid, and in these two operations some fixed air is produced.

(2) Distilled with vitriolic salts, it produces sulphur.

[aa] *The first reaction is a simple reduction to metallic lead and arsenic; the next three are obvious; the fifth seems to refer to the absorption of CO_2 from a separate combustion in sodium or potassium hydroxide which are thereby rendered capable of effervescence on subsequent treatment with acid; and the sixth relates to the formation of ammonium carbonate.*

(3) With vitriolic acid alone, it evolves sulphurous acid gas.

(4) With phosphoric acid, it gives phosphorus.

(5) With the humid caustic alkalis, it renders them effervescent [*on treating with acids*].

(6) Finally with ammoniacal saltpeter, it decomposes the acid, and the volatile alkali that is evolved produces effervescence with acids. [195]

We have repeated and verified most of these experiments, and we did another of which we feel obliged to render an account.

We placed some powdered plumbago on a small saucer in dephlogisticated air, contained under a Priestley apparatus in an inverted glass phial, and we exposed it at the focus of the Tchirnhaus burning glass that belongs to the Academy. The plumbago burned there very slowly, and the combustion gave rise to little deflagrations that scattered a part of the material about. At the end of the experiment, when the elastic fluid contained in the jar had become much less suited to the maintenance of combustion, the plumbago had become converted on its surface into small globules, which reunited as soon as they touched as two similar masses of mercury would have done. We came thereby to form globules that were more than a *ligne* in diameter. Finally, we ceased the operation when the combustion refused to continue for want of dephlogisticated air. Eight days afterward, we found that $\frac{5}{6}$ of the elastic fluid had been absorbed by the water of the apparatus: this was the fixed air that had resulted from the combustion of the carbonaceous part of the plumbago. The other $\frac{1}{6}$ was inflammable like the gas that is evolved when moist charcoal is distilled. This inflammable air resulted from the decomposition of the water brought about by the iron and the charcoal at the end when the dephlogisticated air was too exhausted to maintain its combustion.[bb] As to the globules, we found that they were much harder

[bb] *The description of this experiment shows that there was insufficient oxygen for complete combustion of the plumbago: the gas that was insoluble in water must have contained more carbon monoxide than is usual in water gas. The fused globules that*

than graphite. Their surface was glassy, they left no traces on paper, and they were not at all attracted by a magnet; on digestion in muriatic acid they released a large quantity of iron, and they left a residue like that which gray cast iron and steel ordinarily give under the same circumstances. These globules were therefore only the ferruginous residue that had been [195] calcined and subsequently vitrified by the heat at the focus and that had retained a portion of the unburned plumbago with which it had been in contact.

It follows from all these experiments that it is not fortuitous, as some authors have believed, that the plumbago of which English pencils are made contains about $\frac{1}{10}$ of iron. Without that metal, plumbago, which must be regarded as charcoal saturated with iron, would be nothing other than pure carbonaceous matter. A final point that proves that the iron is present therein in a true state of combination is that plumbago, when it is pure, is not attracted by a magnet.

We believe therefore that we may conclude (1) that plumbago is a substance that we can synthesize [*composer*] and that, in fact, synthesizes itself every day in blast furnaces where gray cast iron is cast and comes to swim on the surface of the molten metal when this metal, in cooling, rejects the excess of charcoal that it cannot retain in solution. In that sort of purging process, the charcoal carries away all the iron that it can retain in its turn, and plumbago is formed;[3] (2) that in cooled cast iron and steel there is possibly

---

remained could hardly have been metallic if they were nonmagnetic, and they were perhaps a fused ferruginous clay which would behave in the manner described.

[3] Iron is perhaps not the only metal with which charcoal has the faculty of combining naturally. One of us (M. Berthollet) had already remarked that a little fixed air is obtained when many metallic substances are detonated. M. de Lassone had also observed (1) that when zinc is calcined with caustic alkali, a little inflammable air is produced, and the alkali becomes effervescent; (2) that when that metal is dissolved in aerated volatile alkali, inflammable air is also evolved, and the solution leaves a black residue.

These results tell us clearly that zinc can contain charcoal; but we wished to assure ourselves of it on our own account and to repeat M. de Lassone's experiment. For that

some combined charcoal, but there is in it also a great quantity that, being released during the cooling, is disseminated in the mass and is not combined. It is not pure charcoal; it is plumbago that has been prevented from collecting at the surface by the rapidity of cooling and the pasty state of the metal.

Thus, gray cast iron and steel, especially when overcemented, cannot be regarded as homogeneous substances; both of them are the result of solutions that have become turbid by a first cooling and that were then hardened by a greater cooling.

The adherence that the iron and charcoal that enter into the composition of plumbago have for each other assures that this substance is not as combustible as is charcoal free from all combination. It requires a higher temperature to burn, and in order to detonate it a greater quantity of saltpeter is necessary than for a similar weight of charcoal. It is not, as M. Scheele thinks, that plumbago contains more phlogiston than charcoal, but, because the combustion of this substance is very difficult, the parts that are not placed in very favorable circumstances in the detonation do not burn. Also, according to the observation of M. Scheele himself, the elastic fluid liberated by the detonation [198] of the plumbago is not pure fixed air, but it still contains a large quantity of dephlogisticated air that has not been used.

---

purpose, we caused two ounces of zinc filings to dissolve in aerated alkali, which yielded $14\frac{1}{2}$ grains of blackish residue. We then detonated the residue with saltpeter, and we obtained from it some fixed air. What remained in the retort, being yellowish green and rimmed with a violet circle at its surface, contained manganese, but we assured ourselves that it also contained [197] iron, by solution in muriatic acid and precipitation as Prussian blue.

Thus the zinc that we used, which appeared quite pure, contained a small quantity of charcoal, of manganese, and of iron.

At present, it is a question of knowing if charcoal must be united with iron in the form of plumbago in order for it to combine with zinc and with some other metals; or even if it can dissolve in these substances without the intermediacy of iron. In the latter case, it would be possible that on leaving the combination, the charcoal might carry along with it a certain portion of the metal, which would constitute as many different plumbagos as there were metals with which the charcoal could combine. Experiment, however, has not yet put us within reach of verifying this conjecture.

Recapitulation

Cast iron must be regarded as a regulus the reduction of which is incomplete, that is to say, a regulus that still conserves a portion of the base of dephlogisticated air. This follows (1) because this metallic substance, in order to dissolve in vitriolic and muriatic acids, evolves less inflammable air, decomposes less water, and absorbs less dephlogisticated air than does soft iron for this purpose, which proves that it already contains a portion of the dephlogisticated air that is necessary for its solution; and (2) because as a result of temperature alone, cast iron, especially when it is gray, refines and whitens without addition and without the contact with air, which could not occur if it did not contain dephlogisticated air to bring about the combustion of the charcoal that renders it gray.

Moreover, cast iron, especially when gray or black, contains charcoal that it has absorbed in its natural state. This is proved (1) by the faculty that it has of cementing soft iron and of transmitting to it enough charcoal to turn it into true steel; (2) by the black residue that is always found at the bottom of its solutions in vitriolic acid, when the solution is done cold — a residue that, like charcoal, dissolves when hot in inflammable air and gives rise to fixed air on combustion. It is to the greater or smaller quantity of carbonaceous matter that cast iron owes the different colors that it displays on its fracture, that may be controlled by varying the proportion of charcoal in the furnace charge.

Cementation steel is nothing other than iron reduced as well as possible, combined moreover with a certain proportion of charcoal in its natural state. The existence of [199] charcoal in steel appears to us to be proved (1) by the increase in weight of iron when it is cemented in pure and degassed charcoal; (2) by the carbonaceous residue that the steel resulting from such cementation leaves at the bottom after solution in acids and that, like the residue from cast iron, dissolves when hot in inflammable air and gives fixed air by its combustion. As to the metallic reduction, that it is pushed farther in cemented steel than in soft iron is proved by the presence of the bubbles that are seen in blister steel and that can only come from the fixed air formed by the combination of the charcoal with the dephlogisticated air that was still in the iron.

Overcemented steel differs from the preceding only by a greater quantity of absorbed carbonaceous matter. This is proved by a greater increase of weight during cementation, by a greater black residue in the solutions, and principally because this quality is given to iron by forcing the circumstances that facilitate cementation, such as temperature and duration.

Perfectly soft iron would be a regulus in the state of greatest purity; but the softest iron of commerce always contains (1) a little charcoal, as is proved by a light black residue after solution; (2) a little dephlogisticated air, which disengaged during cementation, produces fixed air and forms the bubbles that are always encountered in blister steel, even that made from the softest iron. Moreover, the variations that are observed in the volumes of inflammable gas produced on dissolving different wrought irons prove that the metallic reduction is not always carried to the same point.

Finally, charcoal, after having been held in solution by cast iron or steel in the state of fusion and finding itself rejected by the metal at the moment of the cooling, comes out of the combination, while retaining all the iron [200] that can remain united with it. This charcoal saturated with iron is then plumbago, which separates from the metal and, when the cooling is slow, comes to swim at the surface where it can be gathered in its natural state. But when the cooling is rapid and the pasty state of the metal opposes that purging, the rejected plumbago remains disseminated in the mass and communicates to it qualities of steel. So steel, when in the cold state, must be considered as the result of a turbid solution; and the charcoal that it contains, having at first been held in solution and then rejected because of the cooling, is nothing other than finely divided plumbago, dispersed and not combined.

Index

NOTE: *Topics are sometimes listed in modern terminology, and the indexed word will not necessarily be found on the page referred to. Italic numerals refer to illustrations, boldface to the section in this book written by the author in question.*

acid root, see radical acid
acids, action on iron and steel, 98, 181–89, 220–21, 263, 268, 294, 302–3, 307–9, 314–23
   see also spot test
acier de Carmes, 52n, 71n
acier de grain, 53
acier de Motte or *Mondragon*, 53
acier à rose, 52n, 71n
   see also rose steel
Aepinus, F. M. U. T., 252
aerial acid ($=CO_2$), 209–11, 213–14, 230, 239, 254
affinity, chemical, 168, 271, 277
Agricola, Georgius, 22
air, absorbed during rusting of iron, 245
   reaction with molten cast iron, 209–10, 213, 295
   see also dephlogisticated air; fixed air; inflammable air; mephitic gas; nitrous air; vital air
air jet, use in refining iron, 336
Aix-la-Chapelle, 228
Åkerby, iron from, 181, 185, 186, 224, 229
Albertus Magnus, 36
Alchemy, 7, 8, 30, 110
Alessio Piemontese (pseudonym), 4
Alexander the Great, 32
Allerley Matkel (1531), 3
Alling's test for grain size, 100n
alloys, crystal form of, 149
aloes, 18, 19
alum, 16–9

amber, 18, 19
amianthus, 143
Amontons, Guillaume, 89
analyses of iron and steel, 236–38
analytical chemistry, 167–68
antimony (sulphide), 13, 15, 129, 139
anvils, cast iron, 131
   steel, hardening, 59
Archimedes, 149
Aristotelian theory of matter, vi, 22, 110, 282
armor, hardening, 12, 35, 138
arsenic, analysis for, 227
   causing hot shortness, 284
   in iron, 175, 227–28
   poisoning by, 9
   white, 14, 234
Artliche Kunst mancherley weyse Dinten (1531), 3
assaying, 109, 180, 206
   see also analytical chemistry
augers, hardening, 12
Avis aux Ouvriers en Fer sur la Fabrication de l'Acier, 279

Becher, Johann J., 109
bell metal, 13
bells, 17, 140
bend test, 91–4, 102–6, *103*, *105*
Bergman, Torbern, 122, 165–71, **172–255**, 260, 261, 267, 268n, 269, 274, 277, 280, 284, 297, 300–5, 308–9, 314, 318–19, 331, 338, 342, 342n

349

INDEX

Berthollet, Claude Louis, 184, 261, 275–83, **284–348**, 319
Bessemer process, 45, 209–10, 332n
Biringuccio, Vannoccio, 6, 21–4, **25–8**, 31, 42, 324n
blacksmith, art of, 22–4, 44
blast furnace, analogous in effects to nature, 162
  operation of, 172, 328–29
  reactions in, 127–28, 286, 289, 328–30
  scaffolding, 129
  shape of hearth, 130, 130n
  shutdown, 142
blister steel, 51, 292, 298, 327–28
blowholes, 134–35, 145, 156, *157*
blowpipe analysis, 168, 248, 266
Blum, Michael, 3
body of steel, 67, 75, 87–96, 104–6
bole, Armenian, 17, 19
bombs, casting, 134
borax, 14, 15
Boucher, Pauline A., x, 119, 257
Bouchu, Etienne, 119
Bouck va Wondre, T. (1513), 4, 5
Boyle, Robert, 42
Braås (Smolandia), iron and steel from, 183, 185, 196, 199, 201, 202, 203, 221, 225, 227, 233, 235, 238, 246, 250
brass, crystal form, *163*, 164
Brattefors (Värmland), iron and steel from, 182, 185, 220, 224, 229, 237
brazing solder, 15
Brearley, Harry, 77n, 70on
Brescian steelmaking process, 22, 25–8, 324
bronze, patina on, 150
Brown, Harrison, 188n
Burgmans, A., 251
burins, 52

cadmia, 228
caloric, 9, 178, 215–22, 236–38, 262, 282, 293, 304, 308
calorimetry, 219
calx, iron, 331, 333–34
cannon, boring, 135, 191, 191n
  bursting, 134
  casting, 135, 279, 329

cannon balls, defects in, 133–34, 156, *157*, 288
  solidification mechanism, 138–39
Cammerlander, Jacob, 3
carbon, combination with metals, 265, 345–46
  state in iron, 340–46
  see also charcoal; plumbago
Carinthia, 339
Carme Steel, 52–3, 58, 71n
case hardening, 34, 60
  see also cementation
cast iron, *passim*
  blast furnace operation for different types, 141, 322, 328–29
  chilling makes white, 133, 136
  cracking, 135
  crystallization of, 138, 143
  decarburization below melting point, 335, 337–38
  nature of 128–37, 172–73, 236–39, 286–88, 323–47
  reactions with various substances, 192–207
  remelting in foundry, 331
  shrinkage, 136, 287
  sonorousness, 131, 140–41
cast steel, 274n, 296, 339
  see also English steel
Cavendish, Henry, 278
cementation, 35, 51–2, 111–15, 174, 203–5, 214–15, 253–55, 264–65, 291–99, 309–14, 323–24, 338–39
  by molten cast iron, 26–7, 300, 324
  causing increase in weight of iron, 114, 197, 254
ceruse, rust preventative, 39
chafery, 147, 336
Chalup, —., 278, 284
Chalybes, 33
Champagne, steel from, 73
chaos, 143
charcoal, combination, 265
  conversion to plumbago, 266
  qualities requisite for metallurgical uses, 26, 38, 49, 52, 54, 111, 129, 243, 266, 291, 334
  reaction with hydrogen, 319

INDEX

charcoal (*continued*)
  role in blast furnace, 328
  role in cementation, 52, 309–14
  state in iron 325, 340–46
  see also carbon
Charles's law, 316n
chemical composition as key to nature of materials, 276, 285
Chemical Revolution, 276–77, 282
China, bells from, 23n
chisels, steel for, 52
  testing, 75, 95–7
Chisholm, Alexander, 295n
churches, conversion into gun foundries, 279
Clamecy, steel from, 50, 54, 56
cleavage, 143
Clouet, Louis, 278, 280, 284
coal (mineral), use in metallurgical processes, 43, 54, 114, 289, 334
cobalt, 251
coke, use in blast furnace, 334
colcothar, 253
cold-short iron, 45, 173, 183, 184n, 202–3, 206–8, 233, 238, 242–43, 247–49, 284, 290
  phosphorus in, 233, 247–49
color test for metals, 222–25
combustibility of iron, 287
Committee of Public Safety, 259, 279
Conservatoire des Arts et Métiers, Paris, 277
corpuscular theory of matter, 65
corrosion, 150–53, 244–45
Cort, Henry, 167, 334
Courtivron, Gaspard le Compasseur, Marquis de, 121
Couvin (Liège), iron from, 317
cracks, internal, avoiding during hardening, 58, 71–2
Cramer, Johann Andreas, 107–10, **111–17**
Cronstedt, Axel Frederic, 192, 230n
crucible steel; *see* cast steel, English steel
crystal (quartz), 85–6, 138
  (glass), 200
crystal shape, 156–63, *157, 159, 161, 163*
  in alloys intermediate between metals, 149
  cubic, 156, *157*

dendritic, 123, 158, *159*, 286
  related to shape of molecules, 137–38, 149
  rhombic, 156, *157*
  see also fracture
crystallization, of cast iron, 136, 140, 143, 156–63, 286
  of fibrous wrought iron, 290
  fire as solvent for, 162n, 138
  of metals in general, 149
  of salts, 138, 149
  of slag, 147, 162–64, *163*
crystals, hybrid, 138, 149
cutting tools, hardening, 56

Dalesme, André, 95
Dalton, John, 65, 83
damascening, 23
Damascus steel, 28, 31, 38–9, 113, 169, 278
D'Arcet's alloy, 266
Darmstaedter, Ernst, 4
defects in iron and steel, 38, 45, 69–73, 97–8, 242–44
dendrites, *see* crystal shape
Denman, Anne S., x, 275
density of iron in various forms, 236–38
dephlogisticated air (= oxygen), absorbed from acid when iron dissolves, 307, 321
  action on iron, 255, 295
  behavior in cementation process, 313
  behavior in fining process, 333–34
  in cast iron, 262, 306, 306n, 308–9, 325
  in wrought iron, 326–27
dephlogistication (= oxidation) of iron, 247, 249
  see also phlogiston
Desaguliers, J. T., 91n
Descartes, René, 42, 65, 293n
Déscriptions des Arts et Métiers, 64, 71n, 121
dew, quenching steel in, 57, 59
  makes iron hot short, 48
Diderot, Denis, 191n, 259
diffusion, 271
Dijon Academy, 260, 262
Dingelvik (Dalia), iron from, 183, 185, 220, 225, 229
distillation, 37

drawbench, 95
drawplates, steel for, 39, 62
Drey schöner kunstreicher Büchlein (1532), 3
Duhamel, J. P. F. G., 274n
Dulubre, —., 284
Dunckel, Simon, 3

Edelstein, Sidney, 3
Eisenerz, iron from, 223, 226, 317
Eisenman (furnace bear), 267
Ekström, Daniel, 175
electricity, 149, 250
electrochemical equivalents, 187
electuaries, 8
Encyclopédie Méthodique (Panckouke), 259–61
Encyclopédie, ou Dictionnaire Raisonné des Sciences, des Arts et des Métiers (Diderot), 191n, 259, 260
England, science and industry in, 167
English steel, 74, 206, 220, 225, 229, 230, 231, 235, 237, 274n
Ercker, Lazarus, 42
etching, as quality test, 98n
  iron, 4–6, 16–7, 19–23, 31, 38, 43
  marble, 37–8
  resists (grounds) for, 16–7, 38–9
  see also spot test; acids
ethiops (= iron oxide), 251, 263, 303, 333
Euphrates, 32

Fabricius, Georg, 125
fagoting, 36
Ferguson, James, 3
fibrous structure of wrought iron, 148–49, 290
  see also fracture
files, hardening, 12, 23, 34–5, 60–2
  straightening (above M_s?) during quenching, 61
finery process, 44–6, 146, 288–91, 332–38
  see also refinery process
fire, effect on iron, 150–53, 160
fish-hooks, hardening, 35–6
fixed air (CO_2), 320–22
fluxes, 14, 15, 23, 26, 36, 115, 284–85, 294, 328
foaming during solution of iron in acids, 220–21

Foix, steel from, 73
forging, effect on structure of iron, 290, 292
Forsmark, iron from, 182, 184, 185, 220, 224, 225, 229, 237
fossil iron, 145
Fourcroy, Antoine Fançois de, 343
fracture, test for grain size and general quality of iron and steel, 46–55, 76–85, 98, 100–4, *101*, *103*, 146–48
  cast iron, 157, 159, 286
  standards, 80, 98
France, iron ores in, 49–50
Franche-Compté, iron from, 289n
freezing mixtures, 218
Frencken, H. G. T., 4
furnace, Hungarian, 184, 243
  cementation, 112
  reverberatory, 287

Gadolin, Johann, 168, 216n
Gahn, Johan Gottlieb, 168
Galilei, Galileo, 91
Gallicia, 32
gas, inflammable, 266
  mephitic, 266
  see also air
gases, density of, 319–20
  role in cementation, 271
  thermal expansion of, 180, 315–16
Germany, ironmaking process in, 173
  steel from, 52, 57, 71, 71n, 73, 74
gilding, 17–19, 23
glass, use as flux, 15, 56, 206
Gnudi, Martha T., 21
Gobelins tapestry works, 277
gold paint, 17, 18
Gotha, Landesbibliothek, 3
grains, 50, 51, 147–48, 160, *161*
grain size, testing, 76–84, 77n, 82n, 100–3, *101*, *103*
  standards for, 80, 98
Gränjen, iron from, 183, 185, 187, 190, 205, 220, 227, 229, 254
graphite, 288
  see also kish; plumbago
gray cast iron, *passim, especially* 132–43, 132, 236, 286, 337, 347

Grignon, Pierre Clément, 65, 119–24, **125–64**, 168, 180n, 184n, 204n, 226n, 228n, 267, 268, 277, 282
grumillons, 147
Guérigny, iron from, 316, 324
guns, naval, 286
  see also cannon
Guyton de Morveau, Louis Bernard, 259–61, **262–74**, 277

Hall, A. Rupert, 191n
Hällefors (Sudermania), iron from, 183, 185, 186, 191, 192, 193, 220, 224, 229, 331
hammer welding, 51, 54, 78
hammers, hardening, 59
hardening of steel, 9–12, 23, 32–7, 55–62, 60n, 66–7, 113, 292–93
  theory of, 66–7, 282, 293, 293n, 299
hardness testing, indentation, 68, 98–9, 105, 106
  scratch, 85–6
heat of reaction, 216–22, 303
Hergotin, Künigund, 3, 19
Herzog-August Bibliothek, Wolfenbüttel, 3
Hickey, J. Patrick, x, 165
Hjelm, Peter Jacob, 225n, 342
Hooke, Robert, 42n, 91n
hot-short iron, 43, 45, 48–9, 173, 202, 206–8, 231–34, 242–43, 284
  sulphur in, 48–9, 232–33, 284
Hungarian smelting furnace, 184, 243
Hungary, steel from, 52, 54, 59, 71
Husaby (Smolandia), iron from, 183, 185, 186, 188, 190, 196, 207, 220, 225, 229, 234, 237, 238
Hüttenberg, iron from, 317
Huygens, Christian, 89n
hydrocarbon gases, 319–21
hydrochloric acid (?) used to etch marble, 37–8
  see muriatic acid

Iceland spar, 143
impurities in iron, 222–34
India, steel from, 69
inflammable air, 301–3, 307–9, 314–21, 347
  characteristics of, as yielded by different irons, 211–12, 235–36, 242
  evolved by reaction of iron with water, 235
  evolved on solution of iron in acids, 179–92, 234, 269
  formed from charcoal and plumbago, 266, 347
insoluble matter in iron, 228–34
internal friction, 140
iron, *passim*
  formation in nature, 145
  value in different forms, 175
  varieties of (Bergman), 172–78
iron calx (= iron oxide), properties, 244–50
  reaction with cast iron, 192–98
  see also ethiops
iron carbide, identification of, 281
Italy, steel from, 74

Jars, Gabriel, 121, 191n, 274, 280
Jordan, Peter, 3
Jousse, Mathurin, 41–3, **44–62**, 71n, 284n

Karsten, C. J. B., 281
kettles, casting of, 134
Kirwan, Richard, 268n
kish, 141–43, 288
Knight, Gawin, 252n
knives, hardening, 11, 12, 33–45
Kunstbüchlein, 3, 4

lancets, hardening, 33, 34
Lassone, Joseph Marie François de, 345n
latent heat, 216n
Lavoisier, Antoine Laurent, 68, 167, 169, 260, 276, 278, 302n, 306n, 333
lead, 15, 33
  in cast iron, 142, 206
lead bath for hardening, 88–90
lead oxide, reactions with cast iron, 198–99
Lemery, Nicholas, 253
Léon, Pierre, 122n
Leonardo da Vinci, 91n
Leufstad (Roslagia), iron from, 181, 184, 186, 194, 198, 199, 200, 203, 204, 211, 220, 224, 229, 232, 235, 236, 237, 331

lightning, 149
lime, reactions with cast iron, 200–5
Limousin steel, 50, 56, 59, 73
linseed oil, 12, 16–19
litharge, 15
locksmith, 42, 49, 55
Lowitz, J. T., 270
lye, 13, 15

machining test, 97–8
McKie, Douglas, 216n
magnesia, 167n
magnetism, cobalt, 251
   gold, 218
   iron, 149, 249–55
   iron oxide, 221, 251–53, 263
   manganese, 250
magnets, fine-particle, 252n
mail, hardening, 35
malleable cast iron, 64, 152, 268–70
manganese, 176, 208, 222–27, 251, 304
   analytical methods for, 222–23, 226
   removes hot shortness in iron, 208
manganese oxide, reactions with cast iron, 199–200
Mann, James G., 6
Mappae Clavicula, 6
marble, hardening tools for, 36–7
   shaping by chemical attack, 37–8
martial ethiops, *see* ethiops
Mascall, Leonard, 4
mass action, law of, 277
massicot, *see* red lead
mazéage (refining of cast iron without melting), 335
melting iron, by heat of oxidation, 289
mephitic gas (= CO_2), 266, 271
mephitic sulphur (= graphite), 272
mercury poisoning, 9
mercury sublimate, 16
metals, to make malleable, 15
metamorphoses of iron, 125–65
Metcalf test, 77n
meteoric iron, *see* native iron
Méthode de Nomenclature Chimique, 260
Meusnier, J. B. M., 333
Meyer, J. C. F., 247n, 254, 255, 284, 305
Mézières, military academy, 277–78, 280, 284

Mohs, Friedrich, 68, 85n
Mohs's hardness test, 68, 85
molds, 133–35
   dampness in, 134, 145
   gases from, 133
molecular binding, 217
molecules, size of, 156
Monge, Gaspard, 261, 276–83, **285–348**
Montcenis (Burgundy), iron from, 317
Mortimer, Cromwell, 107
Morveau, *see* Guyton de Morveau
Moström, Birgitta, 167n
mottled cast iron, 132, 325
muriatic acid, solution of iron in, 184–85, 188, 263
Musschenbroek, Peter van, 91n, 253

Napoleon Bonaparte, 278
native iron (= meteoric iron), from Germany, 144
   from Senegal, 145
   from Siberia, 182, 185, 188–89, 222, 224, 229
   gives green color (= nickel?) in niter, 224
   sulphur in, 189
natural steel, 174, 296
Needham, Joseph W., 23
Newcastle coal, 114n
Newton, Isaac, 42, 89n, 217
nickel, 224n, 250–51
niter, fused, in color test for metals, 222–25
   in fluxes, 15
nitric acid, action on iron and steel, 98, 179, 185–86, 189, 294, 302–3
   see also spot test
nitrogen, 302
nitrous acid, 179n
nitrous air, 179, 185–86, 234
Nivernais, steel from, 82
Norrberg, iron from, 183, 185–86, 196, 202, 205, 220, 227, 229, 232, 235, 238, 246, 250

ocher, 17
oil quench, 33, 34
ores, iron, 49–50, 126–27, 143, 174–75, 206, 285, 328
   "steel," 113, 115–16

Oriental steel, 28
  see also Damascus steel
Orléans, Duc d', 69
Österby, iron and steel from, 182, 184–87, 189, 190, 196, 205, 220, 224–26, 229, 233, 237, 244, 246, 249
oxygen, 167, 169, 276, 280–81, 306n
  see also dephlogisticated air

Pallas, Peter Simon, 188n
Panckoucke, Charles Joseph, 259
Paracelsus, 109
Partington, James R., 167n, 261n
Pelletier, Bertrand, 342
pencils, graphite, 263, 288, 342, 345
Percy, John, 45n, 335n
Perrault, Claude, 66
Perret, Jean Jacques, 71, 71n, 98n
phlogiston, 109–17, 128–29, 132, 142, 150–53, 176–209, 234, 250, 265, 268, 270, 276, 282, 301–5, 307, 346
  measurement by hydrogen equivalence, 179–86, 189–90
  measurement by reaction with niter, 211
  measurement by reduction of iron oxide, 192–98
  measurement by reduction of lead oxide, 198–99
  measurement by silver equivalence, 187–89
  reducing and congealing forms (Bergman), 176–79, 301–5
  relation to plumbago in irons, 210–14, 238–39, 304
  theory untenable, 307
phosphorus detection in cold-short iron, 233, 247–49
Piedmont, steel, 50–2, 54, 56–7
pin-making, 97
pipe-casting, 329
Platinerie (Liège), iron from, 317
Pliny, 32
plumbago, 175, 197, 203–7, 209–14, 238–39, 262–73, 288, 304–5, 341–48
  amounts in various forms of iron, 209–10, 228
  chemical nature, 265–66, 304, 341–45, 342n
  combustion in oxygen, 344
  large crystals of, 342
  reactions with cast iron, 203–7
  relation to phlogiston in iron, 210–14, 238–39
  synthesis from charcoal in iron, 341–42, 345
pneumatic chemistry, 266, 268–70, 309, 315–22, 334
poisoning, 8–9
porcelain, Réaumur's, 204
Porta, Giovanni Battista della, 30–1, **32–9**
pot-casting, 144
Prado y Tovar, 191n
precipitation of plumbago in iron, 342, 346, 348
Price, Derek, 30
Priestley, Joseph, 260, 268n, 269, 269n
  apparatus for combustion, 344
Probierbüchlein (1524), 6
puddling process, 44–5, 334
pyrites, 126

quenching, see hardening of steel
quenching media, 9–11, 33–7, 55–62
Quintilian, 172
Quist, Bengt Andersson, 183, 185, 220, 225, 229, 237

radical acid, 176, 177, 210, 250, 253
Réaumur, René Antoine Ferchault de, 52, 55n, 63–8, **69–106**, 123, 170, 204n, 215, 253, 280, 282, 293n, 297–300, 311, 317n, 332n
Rechter Gebrauch d'Alchimei (1531), 4
red lead, 16, 17
reducing principle, see phlogiston; sulphurs
refinery process, 144–46, 144n, 288, 327, 335, 335n
  see also finery process
Régnault, Henri Victor, 316n
regulus, iron, 143–44
Rinman, Sven, 169–70, 254, 260, 265, 267, 271, 294, 295n, 321n, 322
Rohault, Jacques, 65, 293n
rolling mill, 42
rolls, treatment of steel for, 339
Romé de l'Isle, J. B. L., 124, 162n
rose steel, 52–4, 58, 71–2, 71n, 100, *101*

rust, protection against, 31, 61, 80
rustic process of iron smelting, 173

Sachs, M., 3
saffron of Mars (=iron hydroxide), 263
sal ammoniac, use in etching, 10, 13, 16, 17, 19, 23
salt, in etching reagents, 10, 12, 13, 15-7, 19
  use in hardening, 56
saltpeter, see niter, 15
sandiver, 15
saws, hardening, 35
Scheele, Karl Wilhelm, 264, 304, 342, 346
Schubert, H. R., 45
screw press, 43
scythes, hardening, 34, 52, 59
secrets, books of, 3n, 4n, 23, 30, 31
Senegal, native iron from, 145
Seric iron (Pliny), 32
shrinkage cavities, 140, 158, 286
Siberia, native iron from, 182, 185, 188-189, 222, 224, 229
Siemens process, 45
siliceous matter in iron, 230-31, 236-39, 239n, 304
silver, applying to iron and other surfaces, 17-9
silver, replacement from solution by iron, 187-89
Sisco, Anneliese G., x, 1, 63, 68
slag, crystals in, 160-62, *163*
  as solvent for crystallization of iron, 147
  study of, 154
Smeaton, W. A., 261n
smelting, reactions during, 284, 328
  see also blast furnace, operation of
Smith, Cyril Stanley, 6, 21, 41, 110n, 119, 165, 257, 275
Société d'Encouragement pour l'Industrie Nationale, 278
Socotra, 18
softening of steel and iron, 9, 9n, 10, 13, 23
  see also tempering
soldering and brazing, 13-5
solid solution, see crystals, hybrid
Sorby, Henry Clifton, 281
Soret steel, 50, 54, 59
Spain, iron from, 32, 48, 54, 59

sparks from hot iron and steel, 27-8, 132, 151, 294, 322-24
specific heat, 216
Spencer, John R., 23
spot test (acid), for iron and steel, 264, 294-95, 311, 322, 323
springs, steel for, 52
  hardening of, 57-8
Stahel und Eysen, von (On steel and iron, 1532), 1-6, *2*, **7-19**
Stahl, Georg Ernst, 109, 133
steel, *passim*
  an alloy like brass, 272
  inferior to iron for chemical preparations, 274
  nature of, 262-74, 347
  overcemented, 307
steeling (steel-facing), 50, 53
steelmaking, Brescian process, 22, 25-8, 324
  by cementation, 25-8, 111-15, 213
  direct or indirect smelting, 115-16, 213-14
Stockenström, A., 183, 228n, 243
stoves, cast-iron, 133
stone-cutting, steel for, 52-3
structure of matter, 217-18
structure of steel, 66, *105*, 106, 281-83
  see also fracture
sulphur, 12, 14, 33, 43, 111, 114-16, 175, 202, 242
  analysis of, in iron, 232-33
  hot shortness of iron related to, 48-9, 232-33, 284
  mephitic (=graphite), 267, 272
  reactions with irons, 131, 140, 141, 150, 245-46
sulphuric acid, see vitriolic acid
sulphurs (reducing principle), 110, 167, 129-35, 139, 142, 150, 152, 253, 267, 282
sulphurs and salts (Réaumur's steel principle), 66, 66n, 297-99
Sweden, science and industry in, 169-70
Swedenborg, Emanuel, 117, 121
Swedish iron, 306, 313, 317, 319, 327, 328, 333
swords, 5, 31, 33, 69, 98n
  hardening of edge only, 36

Tabor, Heinrich, 169
tallow, 11, 33
taps, hardening, 62
tartar, 13, 15, 17, 19, 233
temper colors, 9, 11, 12, 23, 33–7, 57–9, 71–2, 241, 294
tempering, 9, 9n, 10, 23, 27, 33, 35, 36, 55–8, 55n, 60, 293–94
   see also hardening of steel; temper colors
Tennant, Smithson, 266
testing, machines for, 91–4, 91n, 104–6, *105*
testing quality of iron and steel, 44–54, 63–106
Theophrastus, 37
thermochemistry, 215, 216n, 222, 303
Tiemann, Wilhelm A., 281
tin, giving golden color to, 18, 19
Toledo steel, 53n
Tschirnhaus burning glass, 344

Uchatius, Franz, 335n
Ullfors, iron from, 181, 185, 186, 229

Vandermonde, Charles Auguste, 261, 275–79, **284–348**
varnish, 17, 18
Vaucanson, Jacques de, 277
verdigris, 16, 17, 19
vinegar, 11, 16, 17
vital air, *see* dephlogisticated air
vitriol, 16, 17, 193, 193n, 246–48
   white, 6, 16, 17
vitriolic acid, requires water for action on iron, 244

solution of iron in, 181–84, 188, 262, 268, 314–21

Walloon process, 173
Wäsström, Peter, 169
watch springs, testing, 94
   value of, 175
watchmaking, 98
water, reactions with iron, 244
   synthesis of, 278, 307, 322
water pipes, cast-iron, 142
Wedgwood pyrometer, 260
weights and measures, assay system, 180n
   French, 309n
   Swedish, 180n
welding, 15, 36, 70
   see also hammer welding; soldering and brazing
white cast iron, *passim*
   crystals in, 158, *159*
   nature of, 32, 143–44, 286–87, 306
   oxygen in, 306
   production, 129–32, 329
white precipitate from cold-short iron (Bergman), 233–34, 247–49
Williams, Hermann W., Jr., 3, 4, 18
Wilson, Benjamin, 252n
Wolfenbüttel, 3
Wolfsberg, iron from, 317
wood, hardening tools for, 34
wrought iron, *passim*

Zannichelli, Johannes H., 123
zinc, impurities in, 345–46n
   presence in iron, 176, 228

**WITHDRAWN FROM
LIBRARY - RRC**